SEISMIC WAVE PROPAGATION IN REAL MEDIA

MODELI REAL'NYKH SRED I SEISMICHESKIE VOLNOVYE POLYA

МОДЕЛИ РЕАЛЬНЫХ СРЕД И СЕЙСМИЧЕСКИЕ ВОЛНОВЫЕ ПОЛЯ

SEISMIC WAVE PROPAGATION IN REAL MEDIA

Edited by I. S. Berzon

Translated from Russian by
George V. Keller
Colorado School of Mines
Golden, Colorado

$\left(\frac{c}{b}\right)$ CONSULTANTS BUREAU · NEW YORK–LONDON · 1969

ISBN 978-1-4757-0485-3   ISBN 978-1-4757-0483-9 (eBook)
DOI 10.1007/978-1-4757-0483-9

The original Russian text was published for the O. Yu. Shmidt Institute of
the Physics of the Earth by Nauka Press in Moscow in 1967. Two articles
dealing narrowly with Soviet instrumentation (pp. 112-140 of the Russian
edition) have been omitted from the translation.

Library of Congress Catalog Card Number 74-80752

© 1969 Consultants Bureau, New York
A Division of Plenum Publishing Corporation
227 West 17th Street, New York, N. Y. 10011

United Kingdom edition published by Consultants Bureau, London
A Division of Plenum Publishing Company, Ltd.
Donington House, 30 Norfolk Street, London, W.C. 2, England

CONTENTS

CONTENTS

INTRODUCTION

Until recently, the interpretation of data obtained in seismic exploration has been based on comparatively simple representations of the Earth. The most commonly used representation for the Earth has been a set of thick layers, each characterized by a single value for the propagation speed of seismic waves. During the last several years, more complicated representations in the form of thin layers with vertical velocity gradients, as well as homogeneous thin layers have been considered.

New methods for studying propagation speeds in a medium, particularly ultrasonic logging methods, and the results of theoretical and experimental studies of the dynamic characteristics of seismic waves have revealed that a real Earth is considerably more complicated than the simple models accepted in the past. This has led to a need for more realistic representations of the real Earth as a medium through which seismic waves propagate. Because of this, the Department of Seismic Exploration Methods of the Institute of Physics of the Earth of the Academy of Sciences of the USSR has been carrying out both experimental and theoretical studies on the topic "Selection of Physical Representations of Actual Media and the Study of the Corresponding Wave Propagation Effects." Three major subdivisions have been recognized within this overall program:

1. The establishment of a direct relationship between the structure of a real medium and the basic wave-propagation characteristics. This subdivision includes the study of both the nature of so-called seismic boundaries and the various classes of waves associated with these boundaries.

2. The development of a reasonable representation for an actual medium which may be used for interpreting both travel times and wave characteristics. The development of such a model is based on studies such as those described in Part One of this book.

3. The development of criteria for differentiating between various representations for a medium, to create new interpretive methods, with a concomitant improvement in the resolution and accuracy in subdividing the section.

The solution of these problems requires both theoretical and experimental studies, delimited by the following conditions:

1. Experimental studies must provide data about the structure of a real medium, and about the transmission of seismic waves in such media; and so, they must include:

a) detailed measurements of the properties of the medium with ultrasonic and seismic well logging over the entire depth-range of concern in the investigations; and

b) the study of wave propagation (direct-, reflected-, and refracted-wave types) both on the surface and within the medium;

2. Theoretical computations for wave propagation effects must be based on the use of the actual physical properties of the media as determined from the well surveys, and for various physically realistic representations approximating the actual media; and

3. The comparison of the computations with the experimental data must be oriented towards the choice of the most physically probable representation of a medium. This must serve as the basis for differentiating between various representations for a medium using wave propagation effects and for developing a method for the solution of the inverse problem, that of finding the properties of the medium or its representation from data on wave propagation.

The solution of these problems would be tremendously important in seismic exploration. The primary concern will be to increase the resolution of seismic prospecting methods; that is, to increase the detail with which the section can be subdivided and to obtain more complete information about its structure and physical properties. More detailed interpretations would allow the tracing of thin sequences of layers, of variations in the thickness of such sequences, and recognition of specific beds within such a sequence. Such interpretations might also make direct detection of oil or gas pools feasible, but this will require a high degree of resolution in subdividing a series of beds and a high degree of accuracy in determining physical properties.

During 1963 and 1964, the Institute of Physics of the Earth of the Academy of Sciences of the USSR carried out both field research and theoretical studies on the topic "Selection of Physical Representations of Actual Media and the Study of the Corresponding Wave Propagation Effects." Field research was done in the northern part of the Krasnodar Belt.

The primary aspects of the field research and theoretical studies, the results of which are reflected to a greater or lesser extent in this book, were as follows:

1. Detailed study of the section using ultrasonic and seismic well logs.

Detailed velocity logs of the section were obtained for the solution of the physical problems stated above under the following conditions:

a) As a function of the areal changes in the properties of thin layers within a single medium, to determine the lateral persistence of thin layers and to investigate the relationship between variations in wave propagation and changes in the section; and

b) as a function of the type of velocity section and type of medium. Work was done in various parts of the Krasnodar Belt for this purpose.

2. Study of wave propagation effects associated with seismic boundaries, primarily wave reflection. The primary targets of the study were sequences of thin layers with high velocity, located at depths of 280 and 1350 m.

The field technique was designed to record both longitudinal-reflected and longitudinal-refracted waves associated with the sequence of beds under study. A basic assumption was made, that with the mode of operation used, longitudinal waves reflected at normal incidence were recorded, inasmuch as this is the case for which computational techniques based on velocity sections obtained with ultrasonic logging methods have been most completely worked out.

The primary aspect of the investigation in this section of the work is the study of the relationship between the known structure of various sequences of thin layers and the corresponding wave propagation effects. The results would have a bearing on the following important questions:

a) What structure for a seismic boundary results in a strong reflected wave, which can be used as a marker in exploration? Would such boundaries also lead to multiples that would cause difficulty in tracing deeper boundaries?

b) What is the form relationship between waves of various types for a single sequence of thin layers, and what properties of the medium can be determined by using such a relationship?

3. Study of the effect of the upper part of the section on wave propagation. It is apparent that the weathered layer, due to its inhomogeneity and the high attenuation which takes place in it, acts as a filter for waves arriving from below. However, up to the present, no adequately organized study in this respect had been done.

Experimental work was directed to a detailed study of the velocity structure in the upper part of the section and of its filtering properties. This study provided a basis for defining the structure of the upper part of the section and establishing the filter characteristics for waves arriving from below.

4. Study of the pulse form generated by an explosion (the "direct wave"). The study of the incident pulse form is important in carrying out theoretical calculations and in the comparison of such computations with experimental data in constructing a realistic representation of the medium. The study of the source pulse took the following form:

a) the determination of the conditions under which the direct pulse may be used as the incident wave form in theoretical computations; and

b) the exploration of the nature of secondary pulses recorded with the direct pulse.

5. Theoretical calculations. As has been pointed out above, theoretical computations relating to wave propagation in realistic representations of a medium must be based on properties for the medium determined from well surveys. Computations of synthetic seismograms and spectral responses for the complex situation of a laminated medium may be accomplished only with a digital computer. Therefore, programs were developed for computing synthetic seismograms from acoustic well logs for normal incidence as well as programs for computing the spectrum of waves reflected from an assemblage of thin layers for normal and oblique incidence. In addition, approximate methods for computing spectral characteristics which do not require the use of a digital computer were developed. In addition to solution of the direct problem, methods for solution of the inverse problem from the spectrum of a sequence of thin layers were developed.

The present volume contains the results of an analysis of part of the theoretical and experimental work done during 1963 and 1964 by the Institute of Physics of the Earth of the Academy of Sciences of the USSR.

This book consists of three parts. The first part includes work done on instrumentation for ultrasonic logging and the development of statistical methods for analyzing ultrasonic logging data, along with results obtained in logging.

The second part describes the methods used in and some of the results from theoretical computations of the spectral characteristics of a sequence of thin layers, and a new method for their interpretation.

The third part describes the equipment, methods, and analysis of the results of the study of wave propagation in areas where information about velocities was available from well logs.

The authors wish to express their gratitude to workers at the Krasnodar Petroleum Geophysics Trust for their aid in carrying out the experimental work.

PART ONE

RESULTS OF ULTRASONIC WELL LOGGING

EQUIPMENT FOR FULL WAVE FORM
ULTRASONIC LOGGING

V.G. Gratsinskii and E.V. Karus

The solution to seismic problems depends on the availability of detailed data on the elastic properties of the rocks in the survey area as they exist under natural conditions. It would be desirable to use equipment that records the wave response in the well to impulse excitation in analog form. Such a log would permit us not only to determine the corresponding physical properties of the rock with a high degree of accuracy and to make surface/inhole seismic surveys, but also to solve various problems in the propagation of ultrasonic waves in actual media, which might be important in classifying rock types according to combinations of physical properties.

However, in order to record accurate, reproducible data with such a logging system, it would be necessary to use equipment which can record the largest possible number of wave periods, with a high recording speed and with exact placement of the tool in the well bore.

In view of this, and considering the work which has previously been done with ultrasonic logging, the basic requirements for the desired system may be stated as follows:

a) It must have wideband response (5 to 150 kcps);

b) it must have a reasonably wide dynamic range (50 to 60 dB), which will require the use of a gain-calibrated linear amplifier;

c) it must be highly sensitive (with a gain of 10^5 to 10^6, and an input noise level of the order of 1 μV);

d) it must be noise resistant and have gain stability; and

e) it must operate at temperatures up to 80°C and at pressures up to 300 atm.

The model UZKU-3/2-1* ultrasonic logging system, which essentially meets these requirements, was developed at the Institute of Physics of the Earth of the Academy of Sciences of the USSR in 1957. However, several faults in the system became apparent during use: 1) the dynamic range of the downhole amplifier was not adequate; and 2) the equipment had to be operated on a special logging cable, type RKSK-2, which was not commercially available. As a result, following the 1961 field season, the equipment was modified for use with the standard three-con-

---

*E. V. Karus, P. V. Frolov, and V. B. Tsukernik. Described in papers on the topic "Development of a Method of Ultrasonic Pulse Logging for Studying the Elastic Properties and Attenuation in Rocks Penetrated by Wells," Fondi, Inst. of Physics of the Earth, Acad. Sci., USSR, 1959.

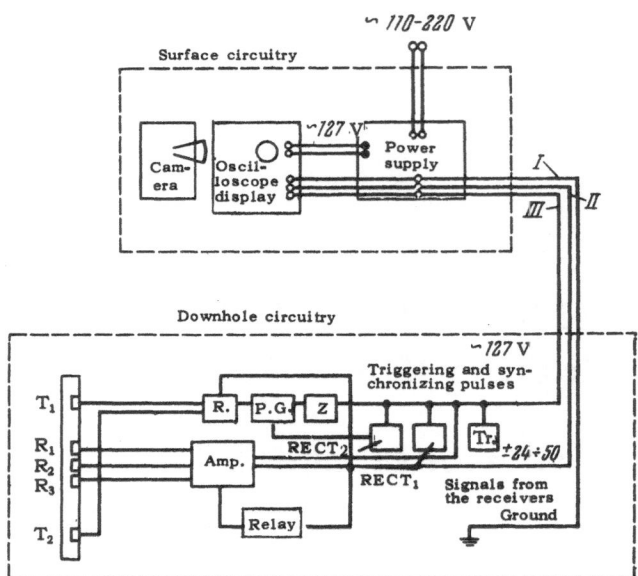

Fig. 1. Block circuit of the UZKU-3/2-2 ultrasonic logging system, with two transmitters and three receivers.

ductor logging cable, with the modification being completed by the beginning of the 1963 field season.*

The modified precise ultrasonic logging system type UZKU-3/2-2† was evaluated during the 1963 field season in the Krasnodar Belt and gave good results.

Use of a cable with only three conductors in place of the RKSK-2 cable which had nine current-carrying conductors (three single and three coaxial pairs) required considerable changes in the circuits between the downhole tool and the surface equipment, as well as some changes in the system for synchronization, downhole power supply and other features in the equipment.

Block Circuit of the Equipment

The UZKU-3/2-2 system consists of surface circuitry and a downhole tool, connected through a three-conductor cable, as shown in Fig. 1.

The downhole tool includes transducers for generating and receiving elastic waves and electronic circuitry for relaying these signals to the surface. As was the case also with the first model, the tool has three receivers and two (or four) transmitters, comprising the measurement system. In the newer model of the tool, it is held against the wall of the well bore by a pressure arm. The signals from the three piezoelectric transducers are amplified by a special amplifier with a commutated input which alternately selects one of the three-input signals (from each of the transducers) and a single output. The four-segment commutating switch allows the signals from each of the three receivers to be sampled sequentially, with samples being transmitted up the cable.

In order to have a wide dynamic range in the amplifier and to avoid overloading the output stage, a relay system is provided for changing the amplifier gain, with the relays being controlled from the surface.

*In this development, B. P. Sinyagovskii took part along with the authors.
†UKZU-3/2-2 indicates an ultrasonic logging system with three receivers and two transmitters, second model.

The electrical pulses for exciting the transducers are formed by the pulse generator, with a specified delay Z introduced in each of the commutator circuits. An intermediate relay R is used to connect the pulse generator either to the upper or to the lower transmitter.

The downhole equipment is powered by ac transmitted over the logging cable at 127 V, 50 cps. The plate voltages for the pulse generator (500 V) and the other circuits (250 V) are provided by separate rectifiers, $RECT_1$ and $RECT_2$, while the heater supply at 6.3 V is provided by the stepdown transformer Tr.

The electrical signals from the piezotransducers are transmitted over the logging cable to the surface equipment, which consists of an oscillograph, camera, and power supplies. A four-segment commutator is used to separate the signals from the various recievers so that they may be recorded on separate tracks. The signals are recorded on three tracks, and the time marks are recorded on the fourth track.

The received wave forms are recorded photographically on 35-mm film. The depth of the inhole tool is also recorded with the wave forms.

The power supply system serves first to provide voltages of various magnitudes and polarities to operate the relays in the downhole tool, and second to provide ac power to the surface and downhole circuits.

The following signals are transmitted over the three conductors in the logging cable: on one conductor, the triggering and synchronizing pulses, along the 127 V ac power; on the second conductor, dc voltages of various polarities and magnitudes ($+ 24$-50 V) are transmitted downward while the signals from the receiving transducers are transmitted uphole; and the third conductor serves as a common ground.

We will now consider the separate components in the UZKU-3/2-2 logging system.

Oscillograph. As was noted above, the oscillograph is essentially the same as that used in the earlier model UZKU-3/2-1 equipment. Details of the circuitry are given in reference [1].

A pulse generator, consisting of the two pentodes stabilized by a quartz crystal, forms negative pulses at a repetition interval of 10 μsec (frequency of 100 kcps). These are used as time marks on the wave form records and to synchronize operations in the rest of the circuitry. These pulses are counted down to two lower frequencies by two blocking oscillators, with countdowns by factors of 10 and 100.

The pulses counted down by a factor of 10 are used to accentuate each tenth timing mark on the oscillograph record.

The pulses counted down by a factor of 100, or to 100 cps, are used for triggering and synchronization both downhole and at the surface. The triggering and synchronizing pulses are coupled with thyratrons. The triggering pulses are of negative polarity, while the synchronization pulses are of positive polarity, and are only one-fourth as abundant as the triggering pulses.

A phantastron delay circuit allows the horizontal sweep to be triggered with some delay with respect to the timing of the transmitted pulses, so that only the usable part of the received signal needs to be recorded. The delay ranges from 60 to 2400 μsec.

A sweep-generator circuit triggered by the trailing edge of the delayed pulses forms a ramp-voltage wave form for horizontal displacement of the light beam, and a positive step voltage for returning the light beam to its initial position.

A four-segment sweep is matched to the commutation scheme by a circuit consisting of two triggers and a mixer which forms a voltage wave form with four steps, used to shift the light beam vertically. The signals from the three respective transducers are recorded on the three upper sweeps, with the signals from the cable being amplified with a two-stage amplifier, followed by an output amplifier stage with paraphase output.

The time marks are recorded on the lowest sweep, and the four-step voltage is fed to another paraphase output stage. The output is connected directly to the deflection plates on the cathode ray tube.

Transmission over the three-conductor logging cable required some changes in the method of providing the trigger and synchronizing pulses to the downhole equipment. Because the number of cable conductors is so much less (3 instead of 9), the solution used was the transmission of negative pulses for triggering and positive pulses for synchronization over a single conductor. With the UZKU-3/2-1 model, each of the synchronizing pulses was transmitted simultaneously with the trigger pulses, using separate cable conductors. It would not be possible to transmit the synchronizing and trigger pulses in such a system over a single wire because they would cancel. In order to avoid this difficulty, the synchronizing pulses are delayed slightly (by 500 μsec). This is accomplished by feeding the positive pulses from the second trigger, which was coupled through a thyrotron in the earlier model, to a bistable circuit (a Schmidt trigger). This circuit provides a negative square pulse with a duration of 500 μsec, and the positive pulse obtained by differentiating the trailing edge ignites the thyratron to provide the synchronizing pulses.

The positive and negative pulses are mixed and are fed to one of the cable conductors through a 0.25 μF capacitor. The 127 V ac power is fed to the downhole circuits over the same wire.

The second active wire in the cable is used to transmit the signals from the piezotransducers to the oscillograph amplifiers at the surface. However, because of the close coupling between the wires in the cable, the 50 cps component and its harmonics are much larger than the signal voltages at the top end of the logging cable. This interference is rejected with a dual T-filter. Rather than the theoretical optimum input impedance of 15 kΩ [2] for matching with the cable, it is found that the best noise rejection is obtained with 30 Ω.

Type KTSh or KTO three-wire cables exhibit high attenuation of ultrasonic frequencies in comparison with that of the coaxial conductors in the RKSK-2 cable. This loss was compensated in the oscillograph amplifier by increasing the amplification and by shunting the cathode resistors of output tubes with 1 ΩF capacitors.

<u>Downhole Circuitry</u>.   The downhole sonde contains the matching amplifier and commutator, the delay circuitry, and the pulse generator and amplifiers for forming the trigger and synchronizing pulses.

The matching amplifier and commutator consist of pentodes. Four-segment commutation is accomplished with two triggers which generate square pulses from the consecutive tubes at specified times, allowing transmission of signals consecutively from each of the three receivers. The fourth segment of the commutator is not used, and time marks are transmitted to the oscillograph during this time.

The delay circuit consists of a triode as a bistable oscillator (a Schmidt trigger). Negative pulses, transmitted over the logging cable and passed through a diode, operate the Schmidt trigger. The positive voltage pulse at one of the plates is differentiated and fed to the grid of a blocking amplifier – pulse shaper. In its normal state, this tube is blocked by the large negative shift, dividing the voltage down by the ratio 1.2 MΩ to 75 kΩ. With the arrival of the positive pulse from the Schmidt trigger (after differentiation), the amplifier conducts, and the negative pulse at the plate is used to trigger the first trigger in the commutator.

A negative pulse of the same duration (about 800 μsec) is obtained from the other plate of the Schmidt trigger circuit, which is differentiated and fed to the control grid of a thyratron for the excitation pulses. In its normal state, the thyratron is blocked by the large negative voltage on its grid, dividing the voltage down by the ratio 1.2 MΩ to 30 kΩ, while a 0.1 μF capacitor at the plate of the thyratron is charged to the plate voltage (500 V). The positive impulses passing through the active commutator segments trigger the thyratron, and the capacitor discharges through a small series resistance, R = 180 Ω, and the internal resistance of the thyratron. The pulses transmitted through the resistance are supplied to one of the piezoelectric crystal transmitters.

Delay of the transmitted pulse with respect to the timing of the trigger pulses is required in order to separate the received signals from crossfeed from the trigger and synchronizing pulses in the cable.

The synchronizing pulses are separated from the combined signals on the logging cable by a second blocking amplifier – pulse former, and then transmitted to a pulse generator. This amplifier blocks with a negative voltage on its grid, dividing the plate voltage down by the ratio 1.2 MΩ to 60 kΩ. A positive pulse causes it to conduct, and a negative synchronizing pulse is transmitted to the plate. In distinction to the case of the trigger pulses, these pulses are fed only to the grid of a single triode, rather than symmetrically to two triggers, permitting square pulses with constant phase relations to control the operation of the amplifiers.

As indicated earlier, to avoid overloading the final stage with high-level acoustic signals, stepwise gain control is provided from the surface. Gain steps in the ratio 1 : 5 : 25 are used, Switching for the desired gain as well as between transmitter elements, is accomplished with three sensitive relays of the type RSM-2.

Plate supply for this circuitry as well as for the pulse generator is provided by special full-wave rectifiers. The 6.3 V supply for the heater is provided by the secondary winding on a stepdown transformer.

Power Supply. The power supply system provides the voltages for the various circuits and for controlling the operation of the downhole circuitry.

Alternating current at 110-127-220 V from a commercial supply or low-power generator driven by a gasoline engine (LS-4) is supplied to the primary of an autotransformer, with 127 V being taken from the secondary to power the oscillograph and other equipment.

Basic Characteristics of the Equipment

1. The maximum amplification in the receiver is 5×10^5, which means that ultrasonic vibrations may be recorded at distances up to 1.5 to 2.0 m from the transmitter.

2. The amplification downhole may be decreased by a factor of 5 or 25 by surface control, so that even large signals may be recorded without distortion.

3. The oscillograph amplifier has a stepwise gain control to reduce the signal strength by factors of 2, 3, 4, 5, 10, 100, and 1000, so that the calibrated dynamic range is about 55 dB.

4. The frequency passband of the equipment is 5 to 150 kcps.

5. The sweep rate and the delay of the horizontal sweep allow recording over the time internal 0-7000 μsec.

6. The pulse repetition rate for the transmitted pulses is continuously variable between 50 and 150 cps.

7. Time marks are recorded each 10 μsec, with each tenth-line being accentuated, so that time differences can be measured with an accuracy of 1.5 to 2 μsec.

8. Scaling of recorded amplitudes over a dynamic range of 60 dB may be done with an accuracy of 10%.

9. An ac power supply (either line or motor generator set) with a capacity of 300 W is required; variations in line voltage of ±10% do not affect the stability of the equipment.

The equipment is designed to operate over the temperature range 20-80°C and at depths up to 2500 m. Some of the design concepts of the UZKU equipment have been used by other organizations in their logging equipment.

LITERATURE CITED

1. P. F. Frolov, Three-channel system for impulse ultrasonic seismic logging, Peredovoi Nauchn.-Tekhn. i Proizvodstvennii Opit., Vol. 8 (1961).
2. D. A. Kokashinskii, Electrical Frequency Filters, Gosenergoizdat, 1959.

UNIVERSAL EQUIPMENT FOR
BOREHOLE ULTRASONIC SURVEYS

L. L. Khudzinskii

Many of the problems which are solved with present-day seismic prospecting methods require that investigations be carried out both on the earth's surface and at considerable depths within the earth (in wells). Well surveys consist mainly of velocity measurements, though in some cases, the attenuation of elastic waves at ultrasonic frequencies is also measured [2, 3, 5]. At present, there are two principal types of ultrasonic (acoustic) logs: a) continuous logs, which permit a detailed study of the variation in velocity of longitudinal waves about a well bore; and b) full wave-form logging, which allows the determination not only of the velocity of longitudinal waves but also of that of transverse waves, and of the attenuation constants α_L and α_T.

The concept of using ultrasonic vibrations for studying the mechanical properties of rocks penetrated by drill holes was suggested in the work of Yu. V. Riznichenko, V. A. Glukhov, E. V. Karus, and V. B. Tsukernik [1, 4]. The accuracy and reliability of the results obtained in such surveys were improved with the development of the new model UZKU-3/2-1 logging equipment. The downhole tool consisted of two transmitters and three receivers, which permitted a high degree of precision in analysis of the measurements.

The UZKU-3/2-1 equipment was designed for use with a 9-conductor logging cable, which severely limited its use. As a consequence, a newer model, the UZKU-3/2-2, was developed under the direction of E. V. Karus, which could be used with the common three-conductor logging cables. The main changes in electronic circuitry were made by B. P. Sinyagovskii and V. G. Gratsinskii, as described earlier in this monograph.

Well surveys are considerably more complicated and costly than surface surveys. The downhole tool must operate at high temperatures (to 100-140°C) and pressures. As a result, the requirements on instrument ruggedness and stability are stringent. Unfortunately, the equipment which has been developed and is in use does not always meet these requirements. For example, the downhole electronics in the UZKU-3/2-2 system do not always exhibit adequate stability, and in a number of cases, the tool has malfunctioned. Because of this, the Krasnodar Expedition of the Institute of Physics of the Earth has developed a modified downhole tool, numbered UZKU-312-2M. This paper is a description of this tool, with emphasis on the changes which had to be made to provide satisfactory downhole service.

Circuit Diagram of the Equipment

The block circuit of the new tool (Fig. 1) has not been changed, inasmuch as it is intended to operate with the existing oscillograph and three-conductor logging cable. The circuit consists of a thyratron pulse generator (1), a relay (2), which switches in the upper or lower transmitter

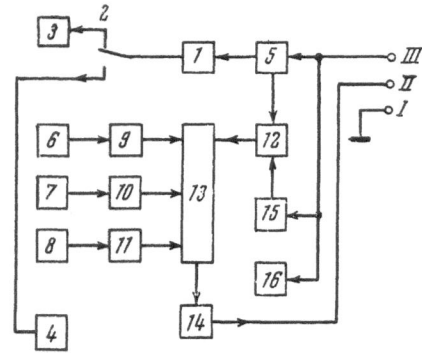

Fig. 1. Block circuit of the UZKU-3/2-2M downhole tool. a) Thyratron pulse generator; 2) relay for sequencing the transmitters; 3) upper transmitter; 4) lower transmitter; 5) time delay block; 6, 7, 8) piezoelectric receivers; 9, 10, 11) preamplifiers; 12) binary counter system; 13) four-segment commutator; 14) output amplifier; 15) synchronization section; 16) power supply.

(3 or 4) alternately, a time delay block (5), three receiving channels (6-11), a binary countdown circuit (12), a four-segment commutator (13), a final amplifier stage (14), a synchronizing pulse selection circuit (15), and a power supply (16).

Triggering of the thyratrons, as well as the action of the commutator and synchronization of the channels is accomplished with synchronizing pulses of various polarities transmitted from the surface. Each full cycle of operation consists of a series of four negative and one positive pulses. The positive pulses are used to fire the thyratrons and in the scale-of-two circuits used in the commutator, but the firing of the thyratrons is accomplished with some delay by using a delay multivibrator. The positive pulses reestablish the synchronization between channels, if it is lost for any reason.

The circuitry is powered by ac at 120 V, transmitted from the surface. The power supply contained in the tool provides the following voltages: a) 6.3 V for the heater supplies of all the tubes except the thyratrons; b) +105 V for plate and screen grid supplies; c) +350 V for the thyratrons; and d) −50 V for providing the bias needed in the system.

Switching of the transmitters, as well as step changes in amplifier gain, is accomplished with relays powered from the surface.

Principles of the Sonde Circuit

Because of the high-temperature environment about the tool and the undesirability of providing cooling for the various components in the tool, components with a high-temperature operating range must be used or common components must be used with downgraded specifications. Inasmuch as the number of heat-resistant components which are available is limited, conventional components are used more or less widely.

Downgraded specifications must be developed in general for electrolytic capacitors, semiconductor diodes, vacuum tubes and resistors. In the case of capacitors, the permissible operating voltage is lower at higher temperatures; in the case of diodes (particularly silicon diodes), the inverse voltage rating and the average forward current are lowered; and resistors must have a higher power rating than that required under normal operating conditions. The amount by which the power dissipation in vacuum tubes can be lowered by lowering the plate and grid voltages is limited, inasmuch as more than 50% of the heat is generated by the heaters. The possibility of using vacuum tubes with cold cathodes in downhole equipment is not presently known.

The circuit contains pulse elements, such as the synchronizing pulse selector, thyratron generator, multivibrators, and triggers. Inasmuch as the tool is being operated with lowered plate voltage for the power supply and thyratrons, in order to increase the power in the transmitted pulse and the stability of operation of the system it is necessary to use as large a part of the available voltage as possible. This is accomplished by using a fixed bias for all of these circuit elements. Elimination of automatic biasing and tubes with grounded cathodes allows increased stability in view of the nearly complete compensation of effects from the filament circuit.

As in the case of the earlier UZKU-3/2-2 tool, this tool is connected to the oscillograph through a three-conductor cable. One conductor of the cable is used as a common ground. The power, triggering pulses, and synchronization are transmitted downhole over conductor III. The voltage for operating the downhole relays is transmitted over conductor II; the signals to be recorded are transmitted uphole over this same conductor.

The power supply for the electronic circuitry is contained in the downhole tool. The transformer has a primary winding, shielding, and three secondary windings. The high-voltage rectifier consists of voltage-doubling circuitry, which provides the best operating conditions for the diodes and electrolytic capacitors for relatively small current requirements. The rectifier supplies for +105 V and −50 V consist of bridge circuits, which have advantages over full wave rectifier circuits with respect to the ratio of average current to peak current from the transformer and with respect to the operating conditions for the diodes. The filaments for all of the tubes are fed from a separate stepdown transformer.

Results of Trials

The electronic circuitry for this tool has been thoroughly tested under field conditions. It was used in logging two stratigraphic test holes and one deep well (well No. 32, 1300-m depth) as part of the experimental program carried out by the expedition.

The UZKU-3/2-2M tool was less sensitive to changes in supply voltage than was the earlier model. Under normal conditions, its operation was not affected by voltage variations over the range from −10% to + 5% of nominal. This characteristic is especially valuable for operations under field conditions, where a stable voltage supply may not always be available. The effect of variations in supply voltage on the stability of operation of the system at high temperatures was not studied.

Principal Shortcomings of the UZKU-3/2-2M Tool and Methods

for Their Correction

1. The low limit for the temperature at which the tool will operate satisfactorily restricts the use of the tool to depths no greater than 2-2.5 km.

2. The large diameter of the tool does not permit its use in small diameter holes, which may be of interest in both oil and mineral exploration.

3. The equipment is designed for operation with a three-wire logging cable, whereas it would be desirable to use it with a single-wire cable.

4. The records obtained with the use of the new system indicate lower resolution and a narrower dynamic range than those obtained with the earlier UZKU-3/2-2 tool. It will be necessary to expand the dynamic range and to widen the frequency band and the filter frequencies in order to obtain quantitative records.

These shortcomings may be compensated for in one way or another. In the UZKU-3/2-2M system, the components with the poorest thermal characteristics are the thyratrons, in which the cutoff characteristics change with temperature, and at 100°C their operation is not compensated. Possibly, a blocking oscillator using vacuum tubes would be more satisfactory as a pulse generator for use in deep holes. The use of transistors or silicon-controlled rectifiers would introduce the hazard of mechanical disintegration on repeated heating and cooling.

A considerable increase in the temperature limit may be achieved if the circuitry is based on the use of cold-cathode tubes. The most promising tubes of this type are the rod-style tubes, which have good electrical and mechanical characteristics, and which require one or two orders of magnitude less power than tubes with indirect heating.

Still another approach to increasing the operational depth for the tool which has not been used is the use of semiconducting thermal devices to transfer heat from the internal circuitry to the outer metallic shell of the tool.

The diameter of the tool is determined first by the size of the transformers. The tool diameter may be decreased in several ways; a) by supplying dc power downhole; b) by supplying ac power at a higher frequency; and c) by reducing the power requirements through the use of cold-cathode tubes.

Conversion to operation with a single-wire cable is possible in principle, but it would require some changes both in the downhole tool and in the oscillograph. However, other approaches to block circuits for ultrasonic logging which require less complicated methods for operation over a single wire are not being ignored.

LITERATURE CITED

1. E. V. Karus and V. B. Tsukernik, Ultrasonic pulse system for studying the mechanical properties of rocks, Izv. Akad. Nauk SSSR, Ser. Geofiz., No. 11 (1958).
2. E. V. Karus and M. V. Saks, Impulse ultrasonic logging, Vestn. Akad. Nauk SSSR, No. 4 (1961).
3. E. V. Karus, R. F. Frolov, and V. B. Tsukernik, Study of the elastic and attenuation properties of rocks penetrated by wells by the ultrasonic logging method, in: Present Status and Prospective Development of Geophysical Methods for Exploration and Development of Mineral Deposits, Gostoptekhizdat, 1961.
4. Yu. V. Riznichenko and V. A. Glukhov, On impulse ultrasonic seismic logging, Izv. Akad. Nauk SSSR, Ser. Geofiz., No. 11 (1965).
5. O. G. Sorokhtin, Multichannel impulse ultrasonic seismic oscillograph, Tr. Akad. Nauk SSSR, No. 6 (1959).

ACOUSTIC INSULATION FOR ULTRASONIC LOGGING*

V.G. Gratsinskii

One of the most important components in a downhole system for ultrasonic logging is the acoustic insulator which is used to prevent signals from propagating along the case of the tool between the transmitters and receivers. A poor-quality insulator would result in the appearance of a false first arrival which would interfere with the wide usage of such logs.

1) The insulator must have a low speed of propagation for elastic waves (at least less than the speed of propagation in water, 1500 m/sec); and

2) the insulator must have a high attenuation factor for elastic waves, so that a maximum attenuation is obtained for waves traveling along the tool.

A number of insulator designs which satisfy these requirements have been described both in the Soviet Union and in other countries [1, 2]. Various approaches have included the use of resin insulators, insulators made in the form of accordian-pleated metal sheets, laminated insulators, and insulators in the form of metal straps with heavy mass filters or energy absorbers, and so on. Nearly all of these insulators have a flexible or semirigid construction which makes it possible to pass through the insulator only a few of the required electrical conductors for transmitting signals between the various parts of the tool. These insulators have the following shortcomings:

1) Semirigid construction allows flexure of the tool in the well, which results in large errors in the measured quantities (for example, the propagation speed for elastic waves); the same effect results from the extension of the insulator section caused by the weight of the lower part of the tool or from compression of the insulator section under high pressure, so that the spacing is changed; in these cases, the errors are systematic in character;

2) pronounced bending of the tool during operation or transport is a source of deterioration of the light-weight wires passing through the inside of the insulator; this causes considerable difficulty and requires special equipment for maintenance and transportation;

3) the need for a special container for the electronic circuitry, which markedly increases the size and weight of the tool; and

4) resin insulators have small mechanical strength under tension, which may lead to the loss of the tool when it becomes hung up in the hole, or when a heavy weight is used at the bottom of the tool.

This paper describes the construction of an acoustic insulator which meets the above requirements and which does not have the disadvantages of existing insulators.

---

*V. G. Gratsinskii. Acoustic insulator. Author's certificate 189164, issued September 13, 1964.

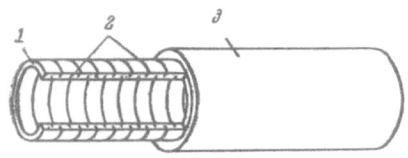

Fig. 1. Construction of the
acoustic insulator.

Description of the Insulator

Because of the nature of ultrasonic logging, the spacing between transmitters and receivers on the inhole tool must be of the order of 1-3 m. The construction of the insulator described below permits the use of nearly all of this space inside the insulator for electronic circuitry or other essential components, whereas this space is almost completely filled with acoustic insulating material in existing tools.

The proposed insulator consists of a metallic tube (1 in Fig. 1) having narrow transverse channels (2) cut to some depth on both the inside and outside walls of the tube along the entire length of the tube. The surface of the tube is covered with a layer of resin or other soft material (3).

In the preparation of the insulator, the channels may be cut in any shape, and may be located in any order (even in the form of a screw thread), and the open end of the tube may be at either end.

Each of the elements used in the insulator has a specific function. The metal tube provides the tool with adequate rigidity and permits the placement of electronic circuitry or other components within it. Each channel in the tube serves as a boundary for reflection and scattering. These cause part of the elastic energy to be reflected and part to be transmitted as a diffracted wave. These processes are repeated many times over along the entire system, causing a marked reduction in wave amplitude.

The resin coating on the tube serves as an energy absorber which attenuates the waves propagating into it from the metal.

Various types of waves may propagate along the insulator; longitudinal, surface, transverse, head, and others. Obviously, the best insulator will provide high attenuation for all of these waves.

It is particularly important that the speed of longitudinal waves be reasonably low.

Experimental Results

Some of the physical phenomena which occur in the insulator have been studied on a model.

The model consisted of an aluminum tube with a length of 35 cm, a diameter of 5 cm, and a wall thickness of 7 mm. Type IPA-59 ultrasonic equipment was used. A Rochelle-salt transmitter transducer was mounted at one end of the model, and a similar receiver was mounted along the surface of the model at intervals of 2 cm.

The traces a and b in Fig. 2 show the effect of the decreased propagation speed of the waves. Figure 2a shows the records obtained by moving the receiver along a plain tube. The first phase shows a speed of propagation of 5580 m/sec. It is apparent that there is practically no attenuation of the wave.

Figure 2b shows the traces recorded along a profile on a similar tube with grooves cut to a depth of 2.5 mm at intervals of 5 mm. In this case, the first phase has a speed of propagation of 2350 m/sec, which is less by a factor of 2.5.

Figure 2c shows traces recorded along a profile on a tube of soft resin. The marked attenuation of all wave phases is particularly notable. The most intense wave phase is attenuated sevenfold at a distance of 30 cm.

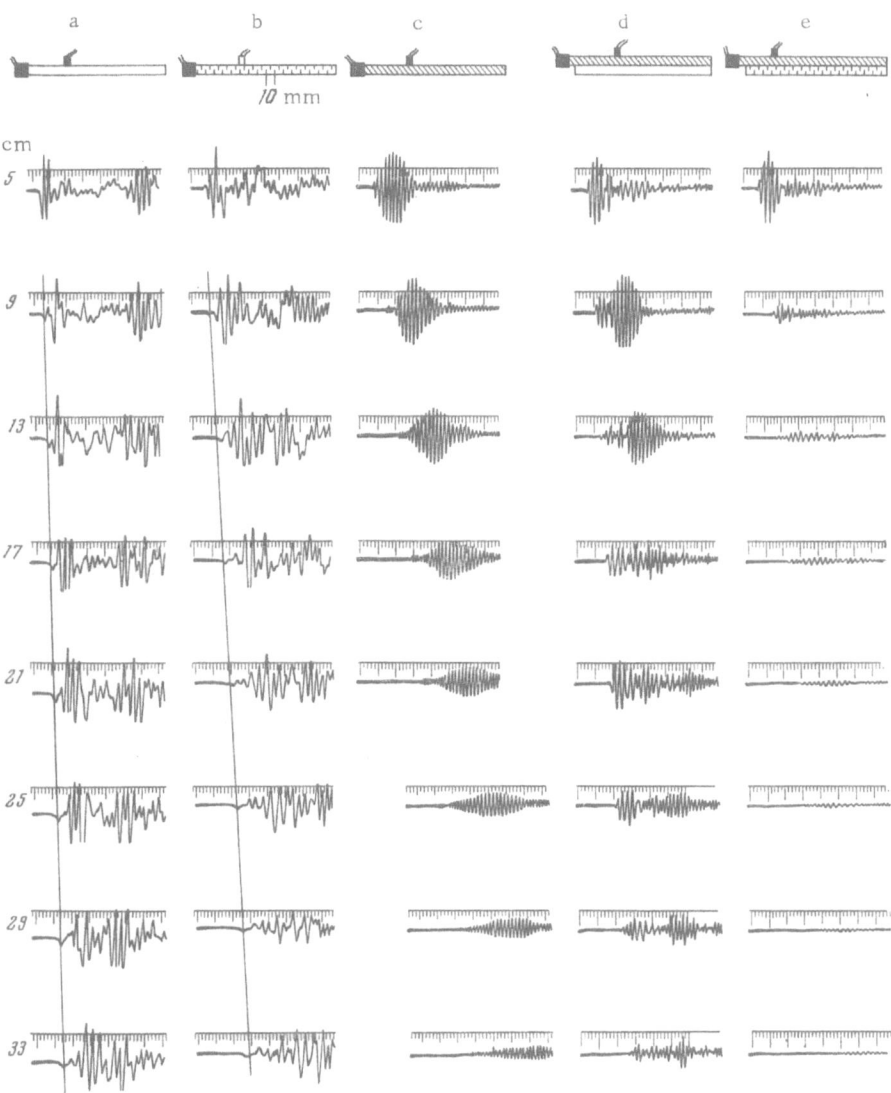

Fig. 2. Modelled acoustic insulator.

Figure 2d is a similar profile for the case in which there is a simple aluminum tube under the resin. Acoustic contact between the resin and the aluminum was accomplished with castor oil. In this case (for similar initial wave amplitudes), the first arrival is a relatively intense high-speed wave propagating through the metal with relatively slow attenuation.

Figure 2e shows a profile in which the resin covers aluminum in which grooves have been cut to a depth of 2.5 mm at intervals of 5 mm. It is apparent that the sevenfold attenuation seen for the resin at a distance of 30 cm (Fig. 2c) is found here at merely 8 cm; that is, the waves are attenuated nearly four times more quickly, and they are nearly absent at a distance of 30 cm.

It should be noted that the traces on profiles c, d, and e were recorded with a gain five times higher than that used for the traces in Fig. 2a and b.

Thus, consideration of resin insulators with and without the supporting tubes shows that one does not obtain the sum of the individual effects, but a completely different effect resulting in high attenuation. This strong attenuation is apparently explained by the fact that in the case of a smooth tube the acoustic energy travels along the boundary and does not pass from the one material to the other, inasmuch as the angle of incidence between the two media is practically 90°. With the grooved metal tube, scattering and diffraction become much more important, with a large part of the energy from the metal leaking into the resin. Therefore, a much larger part of the energy from the metal is absorbed in the resin.

A prototype of such an acoustic insulator was manufactured in the shops of the Institute of Physics of the Earth. The insulator consisted of a steel tube with a length of 1000 mm, and internal diameter of 72 mm, and an external diameter of 96 mm (wall thickness of 12 mm), and with slots cut to a depth of 8 mm at intervals of 10 mm along both the inner and outer walls. The surface of the tube was coated with a resin layer 10 mm thick. The interval volume of the tube, available for placement of electronic circuits or other components, was 4100 cm^3. Figure 3 shows the results of laboratory tests of the insulator. For these tests, a transmitter was located at one end of the device and a receiver at the other end.

Figure 3a is a photograph of the oscilloscope screen of the IPA-59 equipment, with transmitter and receiver located at opposite ends of a simple steel tube with the same dimensions, without grooves or the resin coating. It is obvious that the vibrations are so intense that they do not die out in one repetition interval (20 msec), and the first arrival of the next cycle is lost in the reverberations from the preceeding cycle.

Figure 3b shows the vibrations transmitted along a thin copper wire with a length of 100 cm and a diameter of 1 mm.

Figure 3c and d, shows the vibrations transmitted along an insulator when the transmitter and receiver are mounted directly on the steel body (c) or a resin coating (d).

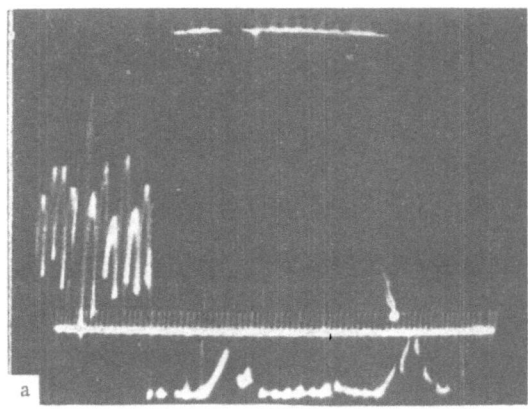
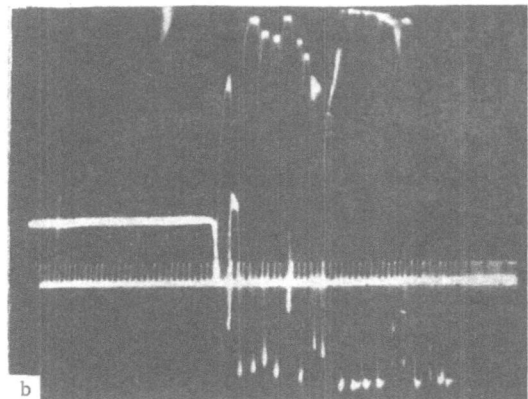
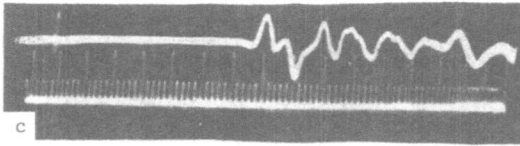
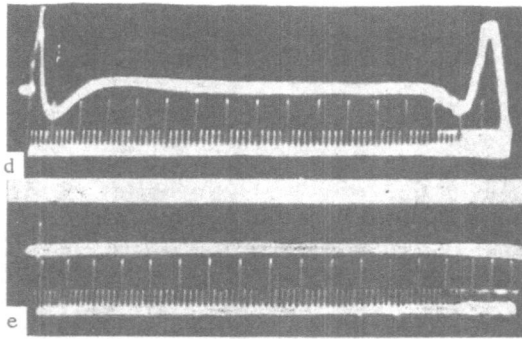

Fig. 3. Seismic traces showing the effectiveness of the prototype of the insulator.

Figure 3e shows the vibrations transmitted along a conductor of the PVR type of the same length as the acoustic insulator. Such conductors are usually required for transmitting electrical signals between the instrument components separated by the insulator.

It is obvious that in the case shown in Fig. 3d, the first energy arrives at 700 μsec. This indicates that the speed of propagation along the tube is 1420 m/sec, or less than the maximum allowable for an insulator. These vibrations are of a very low-frequency type, apparently head waves. It is obvious that under the high pressures in a well these waves will be much less intense. We note that high-frequency longitudinal waves are essentially absent here.

When the transmitter and receiver are mounted on the resin coating (Fig. 3e), as would be the case in acoustic logging, the transmission of both low-frequency and high-frequency waves is absent. With higher amplification, it was established that the vibrations are a hundred-fold lower in amplitude than vibrations transmitted along a 1 mm copper wire, and a thousand-fold lower than that of vibrations transmitted along a simple steel tube.

Thus, we have developed an acoustic insulator which exhibits a high attenuation factor, a low speed of propagation for elastic waves and which provides a large usable interior volume for placement of electronic circuitry.

LITERATURE CITED

1. A. W. Engle and J. L. Casey, Acoustic insulator for acoustical well logging, U.S. Pat. 2,994,398 (August 1, 1961).
2. G. S. Summers and W. Gravley, Exploring unit for acoustic well logging, U.S. Pat. 2,897,478 (July 28, 1959).

STATISTICAL ANALYSIS OF
ULTRASONIC LOGGING DATA

O.K. Kondrat'ev and M.V. Saks

Data from ultrasonic well logging are used to rock type a section on the basis of elastic properties, to study the nature of seismic interfaces and to construct synthetic seismograms [1-4]. Descriptions of equipment, survey methods, and the results of the reduction of ultrasonic logging data have been given in references [5, 6]. However, these references do not contain a statistical evaluation of the accuracy of the data obtained. Such an evaluation is of considerable merit, particularly in the use of ultrasonic logging data for theoretical calculations.

This paper describes a method for and the results of a statistical analysis [7, 8] of the accuracy of determinations of interval velocity obtained from full-wave-form ultrasonic well logs made with the UZKU-3/2-2 equipment developed by the Institute of Physics of the Earth [9].

Characteristics of the Data Obtained in Full-Wave-Form
Ultrasonic Logging

Let us consider the survey technique. Data are obtained at discrete points in the medium, separated at intervals Δh (with $\Delta h = 1-2$ m) along the well bore. The size of Δh is decreased for an increase in the number of layers per depth interval. At each point, data are taken over an interval of 0.5 m with three receivers located at intervals of 0.25 m. Two transmitters are located at a distance of 1 m from either end of the receiver array (Fig. 1).

The interval velocity over one 0.25-m base repeats that which is obtained with the following measurement.

Usually, several repeat surveys (2-6 times) are made in a single well, both with the tool being raised and lowered. It is obvious that the placement of the tool on repeat measurements is made with an error in depth which amounts to $\pm 1-2$ m at a depth of 300 m, and is larger at greater depths.

Errors involved in determining interval velocities in a well bore may be divided into three basic types:

1) Errors related to mismeasurement of the spacing between detectors (changes in the 0.25- and 0.5-m intervals);

2) errors caused by the distribution of data obtained over a spacing of 0.5-m next to an interval of 1-1.5 m of unstudied section; and

3) errors arising from displacements in depth on repeat surveys.

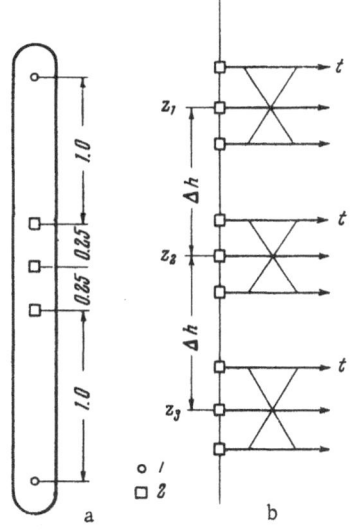

Fig. 1. Survey sytem for ultrasonic logging: a) arrangement of the logging tool: 1) transmitter; 2) receiver; b) system for obtaining data on interval times.

The sizes of errors of the first type depend on the following factors: a) curvature of the tool in the well bore; b) variations in well diameter; c) extent of disaggregation of the wall of the well bore; d) recording of a first arrival of waves travelling through the mud column; e) nature of bedding; and f) inaccuracies in reduction of the data (in picking times, in correlating the proper phases), as well as some others. Errors caused by bending of the tool and variations in well diameter are usually considered to be the essential errors in the data. Except for the last one, the factors listed here are systematic in nature for measurements made at a single point in the medium, but may be considered to be random for measurements made at different points along the well bore. The second and third types of errors depend primarily on velocity inhomogeneities in the medium. Their magnitude increases with increasing inhomogeneity.

For a statistical analysis of the data it is necessary to have a large number of independent measurements of the physical parameter being evaluated — in this case, the interval velocity. However, this is virtually impossible, inasmuch as one cannot assign an exact value for a specific point in the medium, and the tool cannot be repositioned in exactly the same spot for repeat measurements. Moreover, in order to obtain completely independent measurements, it would be necessary to drill a new well, which is not practical.

In this situation there is no possibility of evaluating the precision of each value for velocity, but we may speak about the average error in measurements for a reasonably representative interval in the section. Because in measurement we obtain directly the time difference for the arrival of the first wave at two receivers, we will analyze the mean square error σ_t in determining the time t, which is related to the mean square error σ_V in determining the velocity V, by the expression

$$\frac{\sigma_V}{V} = \frac{\sigma_t}{t}. \tag{1}$$

Brief Description of the Seismic Section in the Survey Area

Below, we will describe the methods and results of a statistical analysis of ultrasonic logging data for the specific case of reduction of data from wells located in the Leningrad field of the Krasnodar Belt. Before going into this, it is desirable to review briefly the nature of the data obtained here and to describe the characteristics of the velocity section. A summary of the data on the velocity section in this area is given by Karus and Saks in the next section of this monograph.

The interval under consideration consists of sandstone and shale beds with rare beds of carbonates. The interval velocity curve has been correlated with the lithologic log of the section. It is possible to recognize several subintervals which may be characterized according to lithology and elastic properties:

1) 115-260 m: a homogeneous sandstone bed, with an interval velocity increasing slowly from 1800 m/sec at the top to 1900 m/sec at the bottom;

2) 30-115, 290-395, and 600-1300 m; interbedded sandstone and shale sequences, with alternating beds of sandstone and shale having a thickness of 1-5 m, and with velocities varying by 200-300 m/sec; and

3) 278-290 and 1348-1382 m; sandstone–shale sequences with thin limestone beds. These parts of the section show the largest differentiation in velocity (the interval velocity for the sandstone and shale is 1900-3100 m/sec, while the interval velocity in the limestone is 2700-5800 m/sec).

It was noted earlier that some of the errors in determining velocities from ultrasonic logs depend on inhomogeneity in the medium. Therefore, the analysis will be carried out separately for each of these subdivisions of the section: for the homogeneous part of the section and for the parts with moderate and pronounced velocity contrasts.

TABLE 1

Well	Log No.	Depth interval, m	Receiver spacing 25 cm				Receiver spacing 50 cm				Tool placement	% Error for 1 m separation
			n	$\bar{y}$, %	σ_y, %	σ_y^0, %	n	$\bar{y}$, %	σ_y, %	σ_y^0, %		
6436	1	15—135					58	—1.6	3.3	3.4	Without clamping	
	2		61	—2.6	4.5	4.9	6	+10.0	5.8	9.1	» »	
	3		11	—0.1	3.5	3.6	25	—1.6	3.4	3.5	» »	
	4		66	—0.7	3.5	3.6	20	—3.0	2.3	3.2	Clamped	
	5		62	—2.6	3.7	4.1	25	—2.7	2.1	2.8	»	
	7		49	—0.8	2.8	3.0	51	—1.6	3.7	3.9	Without clamping	
	Total		251	—1.4	4.4	4.6	185	—1.5	3.5	3.7		
	1	135—261					105	—0.4	3.8	3.8	Without clamping	
	2		67	—2.6	2.8	3.4	39	—3.1	1.8	2.8	» »	
	3						112	—2.2	3.6	3.7	» »	
	4		86	1.7	1.9	2.3					Clamped	
	5						42	—2.9	1.8	2.8	»	
	7						81	+0.5	2.1	2.1	Without clamping	
	Total		153	—0.2	2.7	2.7	379	—1.3	4.4	4.6		1.0
	1	281—296					10	—4.0	10.4	11.0	Without clamping	
	2											
	3						11	—1.2	4.9	5.0	» »	
	4		19	0.5	4.3	4.3					Clamped	5.9
	5						24	—10.1	6.0	9.3	»	
	7										Without clamping	7.9
	Total						45	—6.6	7.2	8.6		7.0
	6	180—185					49	3.6	4.8	5.5	Clamped	
							54	—1.6	5.6	5.7	Without clamping	
39	3	690—700					122	—3.5	6.6	7.1	Clamped	
	3						107	—4.8	13.0	13.5	Without clamping	
	1	1080—1230					75	5.3	7.2	8.1	Without clamping	8.1
	2	630—780					64	4.5	4.4	5.4	» »	

Evaluation of Average Error in Determining Velocity

As a Function of Spacing

At each point on a single log, we have two values for interval time, one obtained for transmission from the upper transmitter t^u and one obtained for transmission from the lower transmitter t^l.

We will take as a null hypothesis that the relative error σ_t/t_{true} in determining times at various points in a section of a uniform type will be characterized by a distribution law which is the same for all of the points.

In this case, the values $y = \dfrac{\Delta t}{\bar{t}} = 2\dfrac{t^l - t^l}{t^l + t^l}$ obtained at these points will be distributed about zero with a variance

$$\sigma_y^2 = \frac{2\sigma_t^2}{\bar{t}^2}\left[1 + \left(\frac{\Delta t}{2\bar{t}}\right)^2\right]. \tag{2}$$

In this expresssion, the square of $\Delta t/2\bar{t}$ may be neglected in comparison with unity, while $\bar{t}$ is approximately equal to t_{true}. Then we have

$$\frac{\sigma_t}{t_{true}} \approx \frac{1}{\sqrt{2}}\sigma_y. \tag{3}$$

In this way, it is possible to obtain the average relative error in determining times for all intervals from single measurements of time. If we assume that t^u and t^l contain only random errors, the average of these two values will be determined with an error $\frac{1}{2}(\sigma_y)$.

Considering that there are a number of systematic errors involved in evaluation of the data, the error obtained in this manner is a lower limit.

In determining the mean square error, a correlation table for values of y was constructed both for individual logs and for the total of all repeat logs, as well as separately for the 25- and 50-cm receiver spacings.

The basic statistics are given in Table 1. The average value $\bar{y}$ is not zero, indicating the existence of systematic errors in t^u and t^l, and moreover that $t^l > t^u$. The nature of these errors has not been explained.

We assume that it is more reasonable to consider an evaluation of the variance of values of y about zero, rather than about $\bar{y}$. We will use such a value σ_y^0 in further discussion.

As may be seen from Table 1, the values for σ_y^0 vary from 2.1-11% for various data intervals.

Errors in Ultrasonic Logging Data Caused by the Lack of Information

on Intermediate Parts of the Section

We have examined the error in determining interval times at observation points (σ_y^0). The total error in determining interval time in a well includes the value σ_y^0 and the error $\sigma_{\delta t}$, related to the distribution of values for these times in adjacent regions where measurements are not made. It is basic that the value for this error will depend on the detail with which the log is made and the degree of inhomogeneity in the section.

For logging of stratigraphic test holes, the tool was advanced one meter at a time, so that the interval skipped was 0.5 m. The interval time for a 1-m section was obtained simply by doubling the time observed over the 0.5-m receiver spacing.

In determining the value $\sigma_{\delta t}$, we assume that we have values for interval time continuously for the entire section, and so for some values $t_{0.25}$ over a 0.25-m interval, we take the value $t_{0.25}$ for the adjacent intervals. In this case, the value for $\sigma_{\delta t}$ may be determined easily from differences in $t_{0.25}$ at adjacent points.

The value for $\sigma_{\delta t}$ may be found not only from all combinations of time differences at adjacent points, $\delta t = t'_{0.25} - t''_{0.25}$, but also from a selection of part of the data which are compiled from point measurements made over some interval. For a large number of such data, the value for $\sigma_{\delta t}$ will approach its mathematical expectation, which will also be the average error for the distribution of measured times in the intervening intervals.

Assuming that the same values $t_{0.25}$ are determined with an error σ_y^0, we obtain the mean square error in determining interval time over a 1-m spacing:

$$\sigma_{y,\,1\mathrm{m}} = \frac{\sqrt{2\sigma(_y^0)^2 + 2\sigma_{\delta t}^2}}{2\sqrt{2}}. \tag{4}$$

Considering the effect of errors in measurement, we can reduce them only by the error σ_y^0. Therefore, with the logging method used at individual points, there is a limit to the precision which may be obtained, given by $\sigma_{\delta t}$.

With the same data, the accuracy in determining velocity (or interval time) is increased if it is determined over larger intervals. Thus, the error in the value for interval velocity for a spacing of H m will be less than the error for 1 m by a factor $\sqrt{H}$.

Values for $\sigma_{y,\,1\mathrm{m}}$ are listed in Table 1 for various parts of the section.

Depth Correlation of Data on Repeated Logging

In making a series of repeat measurements, it is not possible to relocate the tool in exactly the same spots. As a result, the repeat data may vary because of differences in depth of the tool. This error in location in some cases may amount to several meters and be larger than the measurement interval Δh.

In depth-correlating ultrasonic logging data, most commonly the characteristic variations of velocity, which are also correlated with depth, are used. However, there are large depth intervals without such characteristic variations in velocity. Moreover, with such correlations, some points in the medium are given an undue weight.

In some cases, it is preferable to depth-match sets of data by the method of least squares. One of the sets of logging data is taken to have a true depth scale, and the other set is shifted up or down by small amounts Δh. In each case, the departure between individual points on each log is taken and the mean square error for these departures σ^0 is calculated. The minimum value for σ^0 indicates the shift required so that both logs are correlated to the same depths.

An example of such a correlation between two logs is shown in Fig. 2. The minimum value for $\sigma_{\Delta t}^0$ was found for a shift of one of the logs upward by 1 m, and moreover that a shift farther by one interval from the minimum would provide a value for σ^0 which differed from the minimum value $\sigma_{\Delta t}^0$ by an amount less than the permissible error for the final data.

The error in ultrasonic logging on repeat measurements may be determined from the value for $\sigma_{\Delta t}^0$ for the optimum shift. Inasmuch as $\sigma_{\Delta t}^2$ is the variance of time differences for two logs, in order to find the variance for departures from the average it is necessary to divide it by two; thus, the mean square error in determination of time differences is $\sigma_t = \sigma_{\Delta t}/\sqrt{2}$, while the average relative error is $\sigma_y = \sigma_t/\hat{t}$, where $\hat{t}$ is the average value for interval time over the part of the section under consideration.

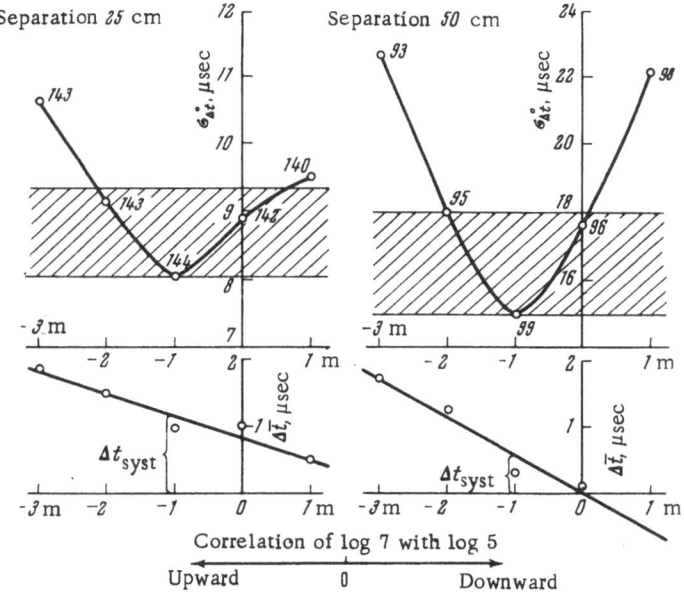

Fig. 2. Correlation of two sets of logging data using the least squares method. The hatched area indicates the minimum range for $\sigma_{\Delta t}^0$; each number at the points on the $\sigma_{\Delta t}^0$ curves is the value for n used in calculating the mean square deviation, $\sigma_{\Delta t}^0$.

Analysis of the results obtained after averaging several logs has indicated that the relative error of averaged data from three repeat logs amounts to $\sigma_y^R \approx 2.9\%$ on the average, while the error for a single log is $\sigma_y^S = 4.2\%$. Thus, the averaged data is more precise, but by a factor less than predicted by theory (a factor of $\sqrt{3}$).

The value σ_y^S does not contain errors related to inaccurate depth correlation, while such errors are included in σ_y^R. Thus, σ_y^R is a sum of relative errors. In view of this, we may assume that the miscorrelation of logs with depth contributes an error of about 2%.

Evaluation of the Effect of Various Factors

On the Precision of Logging

The statistical results have allowed an analysis of the effectiveness of several techniques for running ultrasonic logs and an evaluation of various factors on accuracy. We will make use of the value for σ_y as an objective measure for comparing results obtained under various conditions.

Systematic Errors in Single Measurements. Such errors are explained by the presence of caverns, inhomogeneities in the region immediately around the well bore, bending of the logging tool, and so on. These are avoided by using a reversed survey technique (using the upper and lower transmitters). Therefore, in the case of a significant effect from these factors, the selection of time values determined on repeat logging with a reversed technique must be significantly smaller than for a single run. In one case, it was found that the value for σ_y^R for a one-directional system was only about $\sqrt{2}$ times larger than the value of σ_y^R for a reversed system. This indicates that the differences between t^u and t^l will be largely random. Such results serve as the basis for assuming the validity of the method described above for evaluating the precision of a one-directional survey and allow us to assume that the effect of the factors enumerated above on the precision of logging is negligible.

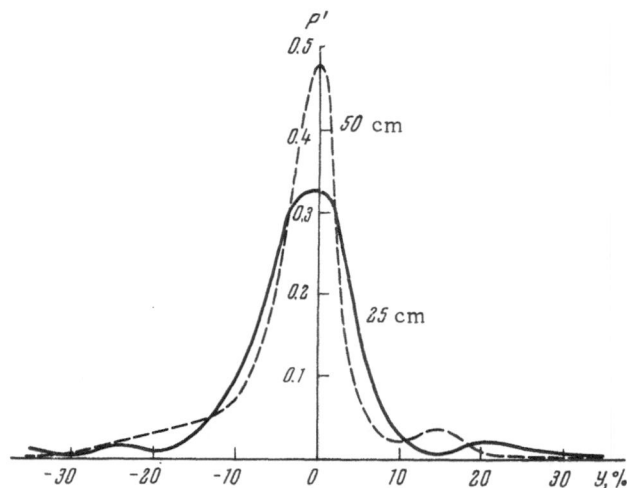

Fig. 3. Distribution curves for relative errors in
interval times for one-directional travel−time curves
for receiver separations of 25 and 50 cm.

Effect of Receiver Spacing. The values of the relative errors in measure-
ment σ_y were determined over the interval 15-135 m with receiver separations of 25 and 50
cm. With the 25 cm separation, a value $\sigma_y = 4.4\%$ was obtained, and with the 50 cm separation,
a value $\sigma_y = 3.5\%$ was obtained (see Table 1). This indicates that the increase in the separation
between the receivers led to an increase in the precision of measurement by a factor of about
$\sqrt{2}$. A similar comparison for the interval 135-261 m (the second log) led to the same result.
This indicates that the separation between receivers has only a slight effect on the accuracy
of the data obtained.

The Effect of Clamping the Tool for Surveys in Holes with Dif-
ferent Diameters. To investigate this, we calculated values for σ_y for various logs and
evaluated the differences between them (see Table 1). It was found that in the case of small-
diameter holes ($d_{well} - d_{tool} = 10$ mm) the difference in values for σ_y between the case of a
clamped tool and an unclamped tool was insignificant. In the case of deep wells with a larger
difference between the tool and well diameters ($d_{well} - d_{tool} = 140$ mm) the values for σ_y found
for clamped and unclamped tools are significantly different. The value for σ_y for an unclamped
tool is larger than the value for σ_y for a clamped tool by 40-50%.

The Effect of Velocity Inhomogeneities in the Medium. A comparison
of values of σ_y for two parts of the section with different velocity structures was made in evaluating the effect of velocity contrasts in the medium. Over the depth interval 110-260 m, the
rocks are nearly homogeneous and uniform in velocity; over the depth interval 273-295 m, the
section includes thin beds with sharp contrasts in velocity.

In the first case, $\sigma_{y, 1m} = 1.0\%$, and in the second, $\sigma_{y, 1m} = 7.0\%$. Considering the difference in these values, they are significant, and as a result, the effect of thin bedding on the error
in determining the velocity from ultrasonic logging data is significant. Thus, the value for the
mean square error is found not for all wells in general, but rather for separate subintervals
with a uniform lithology.

Effect of Propagation of Waves Through the Mud Column. The dis-
tribution curves for values of $\Delta t / \bar{t}$ in the majority of curves are close to normal. There are
differences from such behavior in the upper part of the section (30-135 m), where the distribution is not normal. The departure from normal behavior is characterized to a large degree
by the presence of secondary maximums at large absolute values for $\Delta t / \bar{t}$ (see Fig. 3).

It has been suggested that these are related to the transmission of waves along the mud column. As a consequence, if the excitation from one of the transmitters (the lower one, for example) travels through the rock, while the excitation from the upper transmitter travels through the mud column, the difference $t^u - t^l$ will have a large positive value. With the reversed system, the difference $t^u - t^l$ will have a large negative value. This phenomenon may explain the secondary maximums seen on the distribution curves.

In order to confirm this hypothesis, the pairs of values for t which exhibited the largest values for $|\Delta t/\bar{t}|$ were separated from the paired values for t, and the largest of each pair was taken (max t^u, max t^l). The average $\bar{t}$ for values selected in this way was 165.8 μsec, corresponding to a velocity of 1510 m/sec, close to the propagation speed in the mud.

These values for t were excluded from the distribution curve, with the result that the distribution appeared to be as normal as in the other parts of the section.

Results

The use of statistical methods for evaluating observational errors and the results of determining these errors for measurements in several sections for full-wave-form ultrasonic logging has led to the following.

1. The error in a single measurement of propagation speed for longitudinal waves using ultrasonic logging is ±3-8%.

2. The error in determining formation velocity is determined first by the degree of detail in a survey and the nature of layering in the section being studied. In the case of a homogeneous section, the velocity in a layer 1-m thick can be determined with a precision of ±2.5%, while that in a layer 15-m thick can be determined with a precision of 0.5%. With marked velocity contrasts in the section, the precision may be diminished by a factor of two or three.

3. Average velocity may be determined with a precision of ±1%.

4. In wells with a diameter larger than twice the diameter of the ultrasonic logging tool, the precision is much improved by clamping the tool in the hole.

5. There is a limit to which the precision of full-wave-form ultrasonic logging can be increased, the limit being imposed by the error in the distribution of data at adjacent depth intervals. As a result, a large number of repeat surveys will not increase the precision of the results, and so, it is unnecessary to run more than three logs in a single well.

LITERATURE CITED

1. H. R. Breck and S. W. Schoelhorn, Velocity logging and its geological and geophysical application, Bull. Am. Assoc. Petrol. Geologists, 41(8) (1957).
2. W. J. Van Riel, Synthetic seismograms applied to the seismic investigation of a coal basin, Geophys. Prospecting, 13(1) (1965).
3. J. Delaplanche, Quelques exemples d'utilisation des films synthétiques, Geophys. Prospecting, 9(3) (1961).
4. P. C. Wuenschel, Seismogram synthetic including multiples and transmission coefficients, Geophysics, 30(1) (1960).
5. E. V. Karus, P. F. Frolov, and V. B. Tsukernik, Study of the mechanical properties of rocks by the method of impulse ultrasonic seismic logging, in: Present State and Prospective Development of Exploration Geophysics, Gostoptekhizdat, 1961.
6. P. F. Frolov, Three-channel system for impulse ultrasonic seismic logging, Peredovoi Nauchn.-Tekh. i Proizvodstvennii Opit., Vol. 8 (1961).
7. A. K. Mitropol'skii, Techniques of Statistical Computations, Fizmatgiz, 1961.
8. V. I. Romanovskii, Elementary Course in Mathematical Statistics, Gosplanizdat, 1939.

VELOCITY STRUCTURE IN THE NORTHERN PARTS
OF THE KRASNODAR BELT ON THE BASIS OF
ULTRASONIC LOGGING DATA

E.V. Karus and M.V. Saks

The quantitative relationship between seismic wave processes and the structure in an actual medium may be established only when there is complete data on the mechanical properties of the rocks comprising the section under study, and primarily on the speed of propagation for compressional waves and the density of the rocks.

One of the most important characteristics of a section is the presence of seismic interfaces; their nature determines the relative intensities and other propagation properties for the various types of seismic waves. Therefore, the way in which velocity changes across such boundaries is of considerable interest.

Low-frequency seismic logging has been used in studying the velocity distribution in the Krasnodar Belt. Summary descriptions of the seismic logging data may be found in a number of reports of the Krasnodar Petroleum Geophysics Trust.* Comparison of vertical travel-time curves obtained from seismic borehole surveys run in different wells located in a single field indicates that reproducibility is reasonably good. However, there is a significant scatter in determinations of bed thickness, measurements of formation velocity, and evaluation of the depths to interfaces between layers with different propagation speeds for seismic waves. This largely is caused by insufficient accuracy in the standard methods for reducing seismic well surveys for the conditions encountered in the Krasnodar Belt.

Some survey techniques used in seismic well surveys, such as the use of multichannel recording, reduction of the distance between successive measurement-points in the well and multiple repetition of measurements lead to improved accuracy, which in some cases permits the identification of quite thin beds within the section. However, this taxes the best capabilities of the seismic well surveying method.

Based on surface seismic surveys and electrical well logs, it may be supposed that the upper part of the geological section in the Krasnodar Belt can be characterized by the following velocity structure: the rocks are poorly differentiated on the basis of propagation speeds for seismic waves; there is a velocity gradient (a change in velocity with depth); and there are thin layers with thicknesses that vary laterally. Basically, it would be difficult to study such a sequence with seismic well surveying methods. Knowledge of the characteristics of these thin layers is absolutely essential, inasmuch as they are the main targets for surface seismic surveys in the northern Krasnodar Belt.

---

*Report of seismic party No. 23/63. Internal report of the Krasnodar Petroleum Geophysics Trust, 1963.

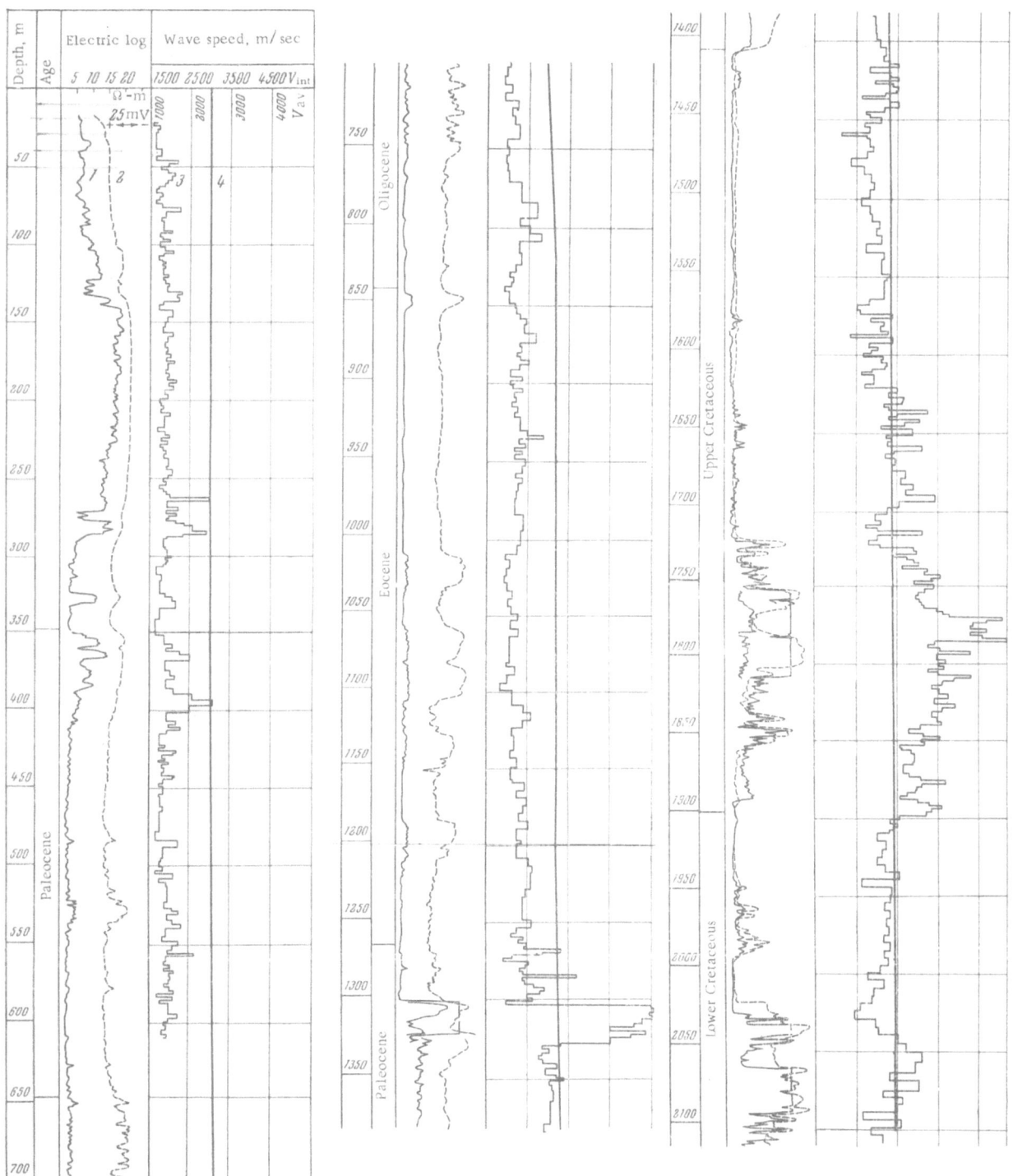

Fig. 1. Ultrasonic logging curves from well No. 52 of the Leningrad field. 1) Apparent resistivity log; 2) self-potential curve; 3) curve for interval velocities from the ultrasonic logs; 4) curve for average velocities from seismic well surveys.

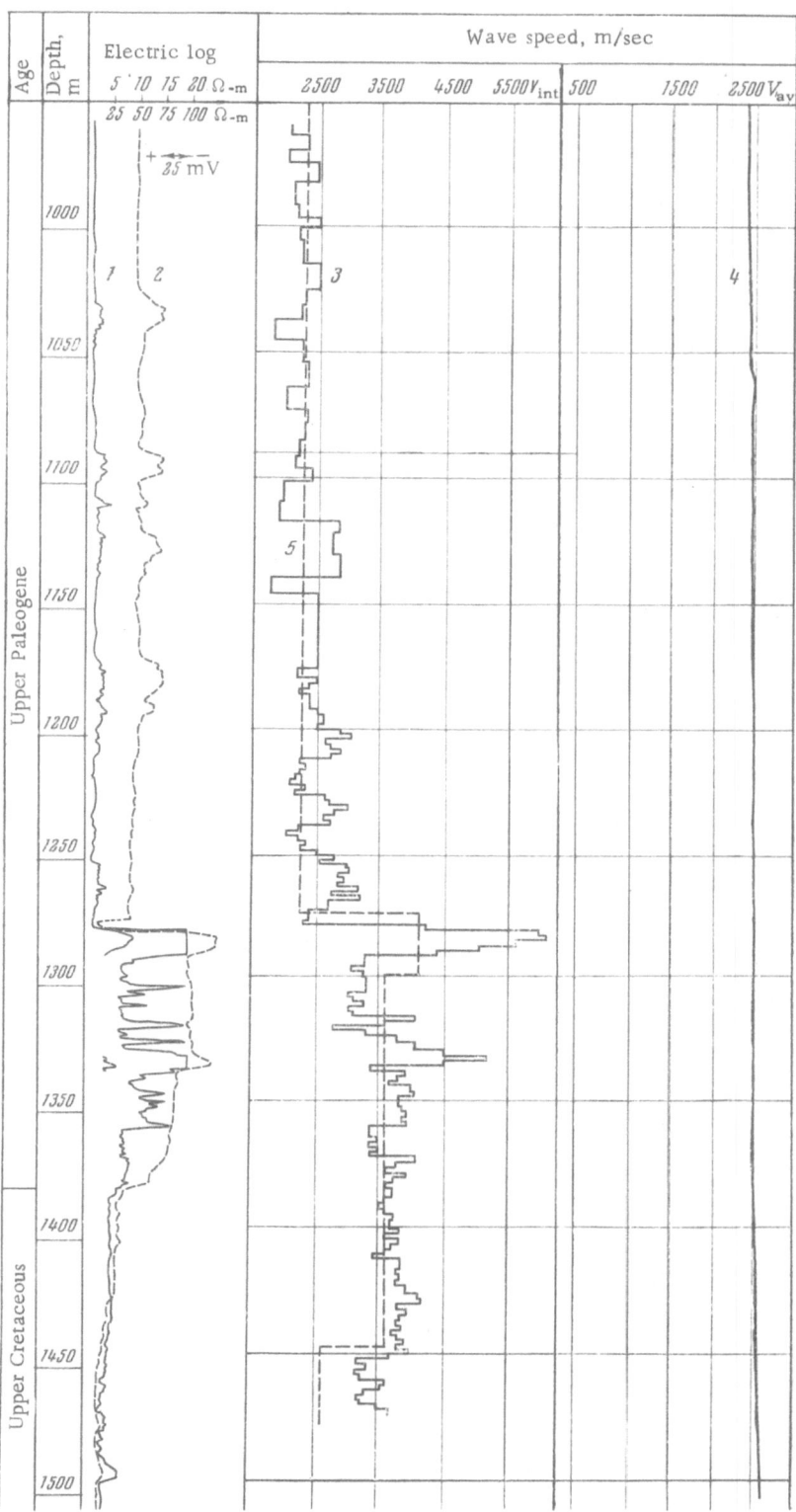

Fig. 2. Ultrasonic logs from well No. 45 in the Staro-Minsk
field. (Legend as in Fig. 1).

The UKZU-3/2-2 ultrasonic logging equipment described by Gratsinskii and Karus (this book, p. 7) was used in the detailed study of longitudinal velocities. A summary of the method and results of work carried out may be found in reports from the years 1963 and 1964.* The measurements were made by V. G. Gratsinskii and K. V. Nekrasov. The ultrasonic logging method was used in the Leningrad, Staro-Minsk, Kushchev, and Krilov fields and in exploration and development wells with depths up to 2100 m as well as in special stratigraphic test holes drilled to depths up to 400 m.

The ultrasonic logging data were used in the solution of the following problems:

a) Obtaining detailed data on the variation of velocity with depth;

b) studying the correlation of different parts of the section over the area, for establishing the relationship with the character of wave propagation obtained with surface surveys, as well as for comparison of the seismic sections in different areas with the purpose of obtaining data on the structure of various types of media; and

c) that of comparing the velocity characteristics with the lithology of the section penetrated by a well.

Nature of the Geological Structure

The northern part of the Krasnodar Belt is one of the boundaries of the Kubano—Black Sea Depression, its northern limb. The sequence in this part of the depression consists of the following rock members:

1. Precambrian basement, which is penetrated by wells only in the northernmost part of the area (the Kushchev field); it consists of strongly faulted granite—gneisses and granites. The upper layer is a weathered zone, its surface strongly eroded;

2. an intermediate sequence, Jurassic and Triassic in age. It is complicated by faulted metamorphosed rocks (slate, claystone, sandstone and in some cases limestones);

3. lower Cretaceous beds (the Albian) are clastic detrital rocks: sandstones, sands, and siltstones, as well as shales with interbedded limestones and marls;

4. upper Cretaceous beds are represented largely by limestones and carbonate-rich dense shales;

5. a thick sequence of Tertiary rocks, consisting of shales, sands, sandstones, and siltstones. In some cases, beds and thin layers of limestones are found, among which the dense limestone in the lower Paleogene should be noted;

6. quaternary clays and bedded sands finish the section.

It should be mentioned that in going south, the thickness of the sedimentary sequence thickens in places.

More complete data about the stratigraphy and structure of the region may be found in internal reports of the Krasnodar Petroleum Geophysics Trust.

Velocity Characteristics of the Various Areas

Leningrad Field. In this area, ultrasonic logs were run in two deep wells (wells No. 52 and 39) and in two stratigraphic test holes. Interval velocities were obtained for depths

---

*Reports for 1963-4 and 1964-5 on the theme: Development of physical representations of actual media and study of the corresponding wave propagation effects. Reports of the Institute of Physics of the Earth, Academy of Sciences of the USSR.

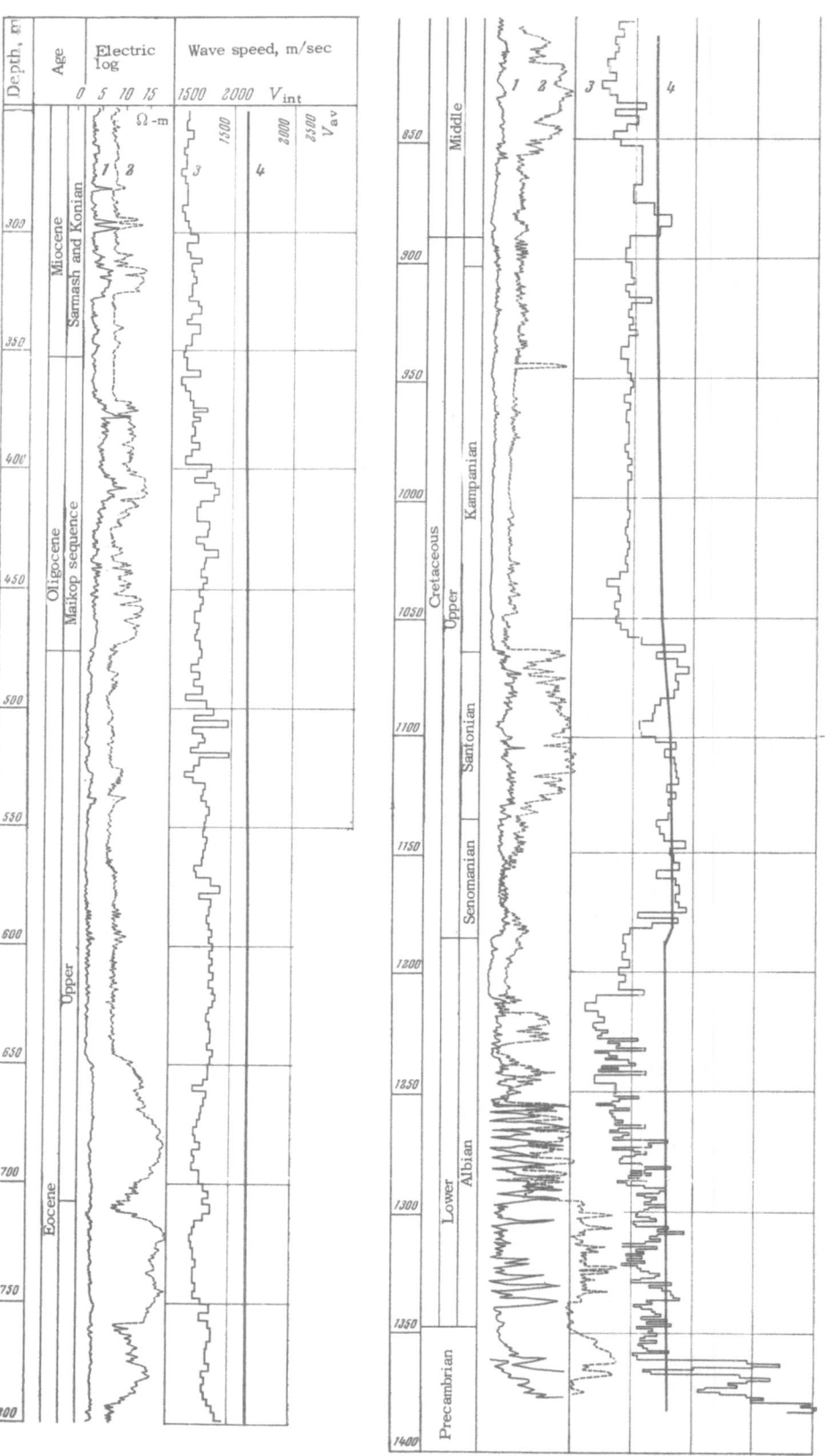

Fig. 3. Ultrasonic well log from well No. 32 of the Kushchev field. (Legend as in Fig. 1).

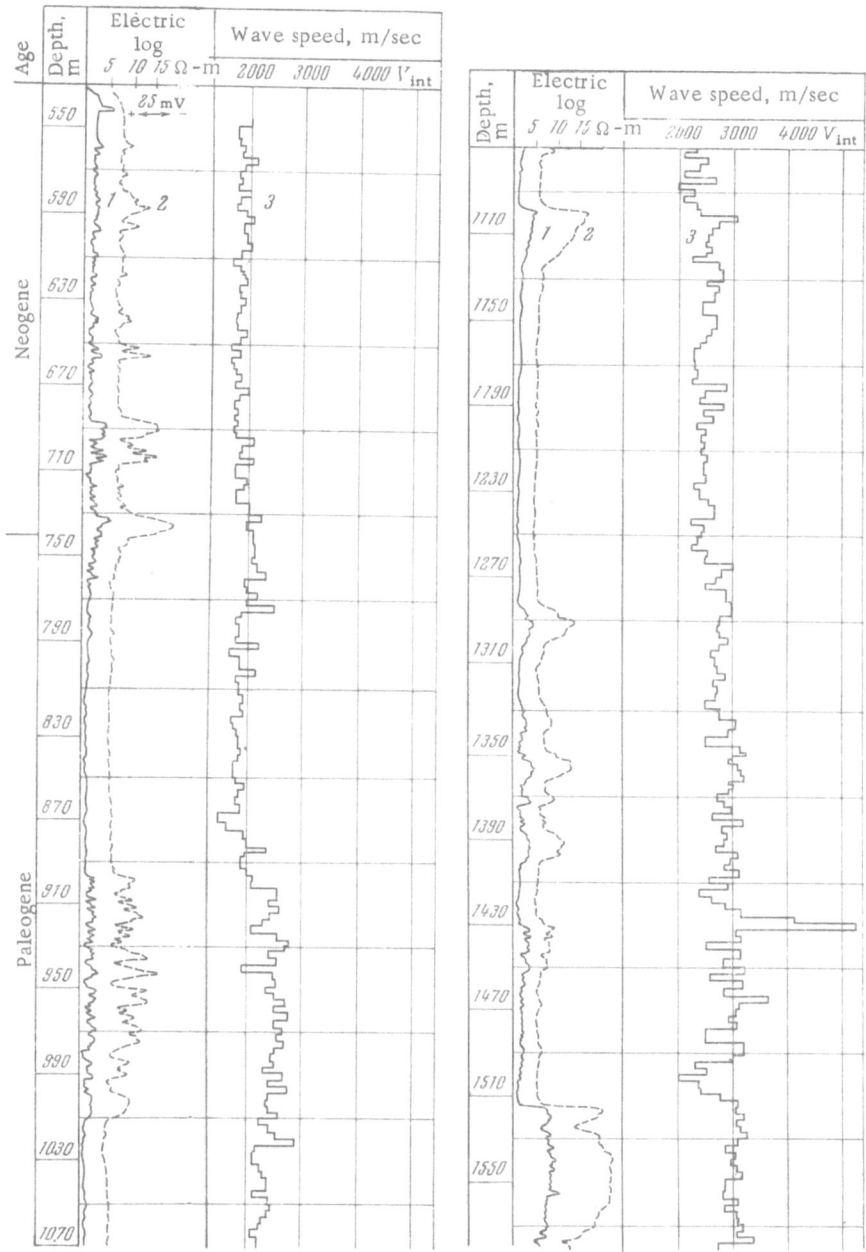

Fig. 4. Ultrasonic well log from well No. 16 of the Krilov field.
(Legend as in Fig. 1).

of 30-2100 m. Curves for the variation of interval velocity with depth in this field are given in Fig. 1.

The diagnostic features of the section penetrated by wells in the Leningrad field are:

1. The existence of a thin (10-40 m) assemblage of layers with high velocity; the nature of the assemblages varies from well to well — they vary in respect to the number of thin beds present and also to the interval velocities within them, but the average formation velocity in an assemblage of layers remains nearly constant over the field; in well No. 52 the tops of the as-

TABLE 1

Depth, m	Average wave speed by area, km/sec					
	Leningrad		Staro-Minsk		Kushchev	
	ultrasonic	seismic	ultrasonic	seismic	ultrasonic	
100	1900	—	1320 *	1320 * 1650	1720	1720
1000	2050	2100	1920	1930	2000	1950—2040
2000	2350	2420	—	2330	—	—

semblage lie at depths of 280 m (with a range of formation velocities of 2.4-2.7 km/sec), of 370 m (V≈2.4-2.5 km/sec), of 1305 m (V≈4.5-5.0 km/sec), and of 1770 m (V≈5.5 km/sec).

2. the presence of layers with high velocity but with gradational boundaries (for example, the bed with a top of 2030 m);

3. the presence of sequences of beds with a positive velocity gradient: over the range 30-140 m and 750-1300 m, the gradient is slight, about 0.2-0.4 km/sec/km, but the gradient in the section from 1500-1700 meters reaches 3.0 km/sec/km; and

4. a sequence of beds with a velocity inversion at depths between 1780 and 1900 m.

The primary targets in surface seismic surveys in this area are the depth intervals 270-280 and 1305-1340 m. We will consider the continuity of these beds over the region later.

Staro-Minsk Field: The geological structure in this area is generally similar to that in the Leningrad field. Ultrasonic well surveys were run in two deep wells (wells No. 45 and 46, Fig. 2) and two stratigraphic test holes (well 6436 and III) over the depth interval 30-1450 m, except that the intervals 400-650 and 800-950 m were not logged. We will consider only the diagnostic differences between the data from the Staro-Minsk field and the Leningrad field in this section.

The layer with high wave speed found in the Leningrad field at depths of 270-280 m has nearly the same properties at well 6436 at Staro-Minsk, but is absent in well III.

Significant differences are apparent in the structure of the Paleogene beds over the depth range 1280-1390 m (Staro-Minsk field) and 1305-1410 m (Leningrad field). At the Staro-Minsk field, the sequence over this interval shows very strong contrasts in wave speed (ranging from 2.5 to 5.7 km/sec). Two thin beds may be recognized within the sequence (H = 12 m and 8 m) with wave speeds of 5.4 and 4.6 km/sec; the tops are at depths of 1285 and 1334 m. The average value for wave speed over the whole section is 3.7-3.8 km/sec, while at the Leningrad field, it is 2.8-3.0 km/sec.

The underlying rocks have wave speeds of 3.6-3.9 km/sec.

Kushchev Field: The velocity profile has been obtained from acoustic logs run in two deep wells (wells 28 and 32, Fig. 3) and one stratigraphic test hole (well V). The properties of the velocity profile in the Kushchev field, based on acoustic logging data, are as follows:

1. An absence of thin beds with sharp, strong boundaries; beds (thickness of 12-15 m) with wave speed higher than that of the adjacent rocks by 0.8-0.9 km/sec are found only at depths of 850 to 900 m;

2. the presence of a thin bed with high velocity, with its top at 1060 m depth;

3. a strong boundary with a contrast in wave speed of 2.0 km/sec at the surface of the crystalline basement; a transitional layer above this boundary may represent weathering on the basement surface; and

4. comparatively weakly differentiated rocks in the upper part of the section, and the presence of a thin-bedded sequence at a depth of 1230-1360 m; there is a positive velocity gradient in this sequence.

Krilov Field: An acoustic log was run in one well (No. 16) over the depth interval 540-1580 m (Fig. 4). Thus, the velocity profile was not fully defined. We note here only the principal features of the logs:

1. In general, velocity increases with depth over the logged interval 1.7-3.0 km/sec;

2. a high-velocity layer with a thickness of 10-12 m occurs at 1430-m depth;

3. a low-velocity layer with a thickness of 20 m occurs at 1495-m depth;

4. there are areas with gradual increases in velocity with depth (690-770 m, 900-1010 m, and 1000-1170 m); and

5. the velocity profile has a thin-bedded character, but the degree of differentiation in velocity is less than that in the sections at Staro-Minsk and Leningrad.

Relation of Average Wave Speed to Depth: The relationship between average wave speed and depth was evaluated using the ultrasonic logging data from the various fields. Average wave speeds from sonic and ultrasonic logs are given in Table 1.

Uniformity of Velocity Profiles by Area

In using ultrasonic logging data in the interpretation of seismic exploration data, it is necessary to determine how well such data represent the general character of an area. Inasmuch as ultrasonic logs have been run in only a few wells up to the present time, we have made use of ultrasonic logging data in conjunction with electric logs in a study of this problem.

Relationships between elastic and electrical properties of sedimentary rocks have been noted in the literature [1, 2]. Velocity profiles were determined from electrical parameters, the apparent resistivity in particular, using such a relationship [3, 4]. In a number of cases, the apparent resistivity curve has been used in computing synthetic seismograms, and good results have been obtained [5, 6].

We have compared sonic and ultrasonic logs from the same wells, and the correlation coefficient for many beds is 0.7-0.85.

A qualitative comparison of ultrasonic logs and electric logs is given in Figs. 1, 2, 3, and 4, from which it is obvious that the basic forms of the two curves are similar. It is possible to estimate bed thickness, the degree of interbedding and the location of boundaries using the apparent resistivity curve.

Figure 5 shows a comparison of ultrasonic logging curves and the apparent resistivity curves for the beds which comprise the target of this study in the upper part of the section (at depths of 200-300 m). The basic characteristics of the thin beds with high wave speeds, as noted on the ultrasonic logs in the four wells in the Leningrad and Staro-Minsk fields, are similar. The similarity in characteristics is substantiated by the electric logs. This layer is not seen on the ultrasonic log from well III of the Staro-Minsk field, but it is present on the electric log.

Another, and the primary target of the study is the sequence of beds with high wave speeds in the Paleocene section (1250-1400 m in Fig. 6). The general characteristics and their comparison with those in other fields are given above.

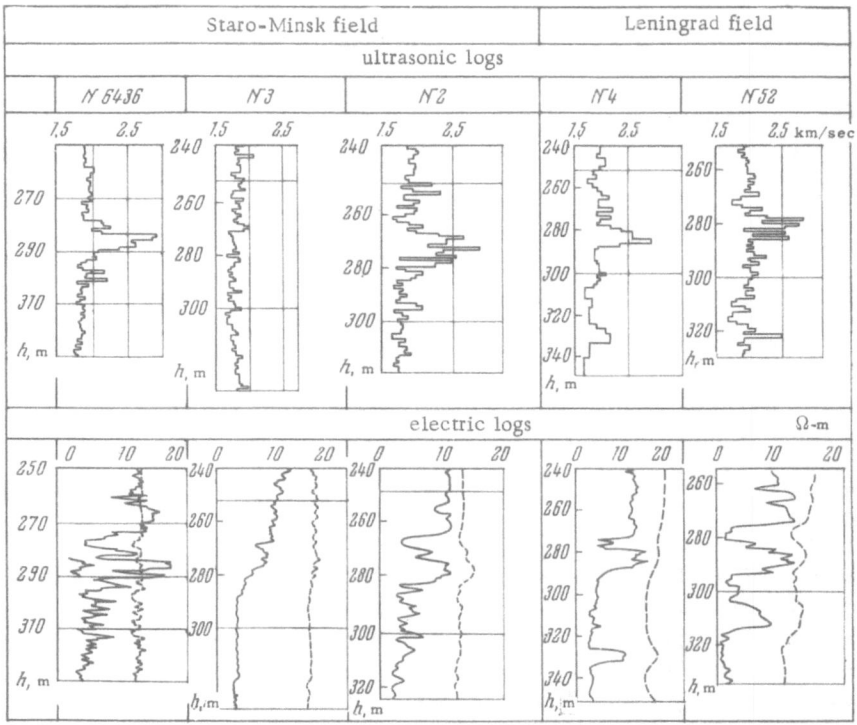

Fig. 5. Ultrasonic and electric logs of the upper part of the section
in the Staro-Minsk and Leningrad fields.

We note here only that the basic characteristics of this part of the section correlate well
with respect to interbedding on the electric logs. The large differentiation of the section in the
Staro-Minsk field, characteristic for well 45, is observed over the entire profile of wells. In
the Leningrad field, the thickness of the high-speed layer decreases from 50 m to 20 m along
the well section from the southwest to the northeast, but the character remains the same.

A similar comparison for upper Cretaceous beds from the Kushchev field is shown in
Fig. 7.

These well sections indicate the uniformity of the basic features of the structure over
these fields.

Wave Speeds for Various Rock Types

The ultrasonic logs have shown that various types of rocks may be grouped according to
longitudinal wave speeds. In so doing, wave speeds were read from the logs over depth intervals corresponding to specific rock types. The logs were averaged over these intervals to determine the wave speed for a rock.

All of these data are listed in Table 2, grouped according to rock type (limestone, marl,
sandstone, sand, or shale).

The precision in determining V_P depends on bed thickness and the depth at which measurements are made, and it may vary over rather wide limits [7]. In a number of cases, rocks
of various types may be interbedded, and this is reflected in the character of the acoustic log.
As a result, the precision in determining V_P is not high. The data for wave speeds for various
rock types are given in Table 2.

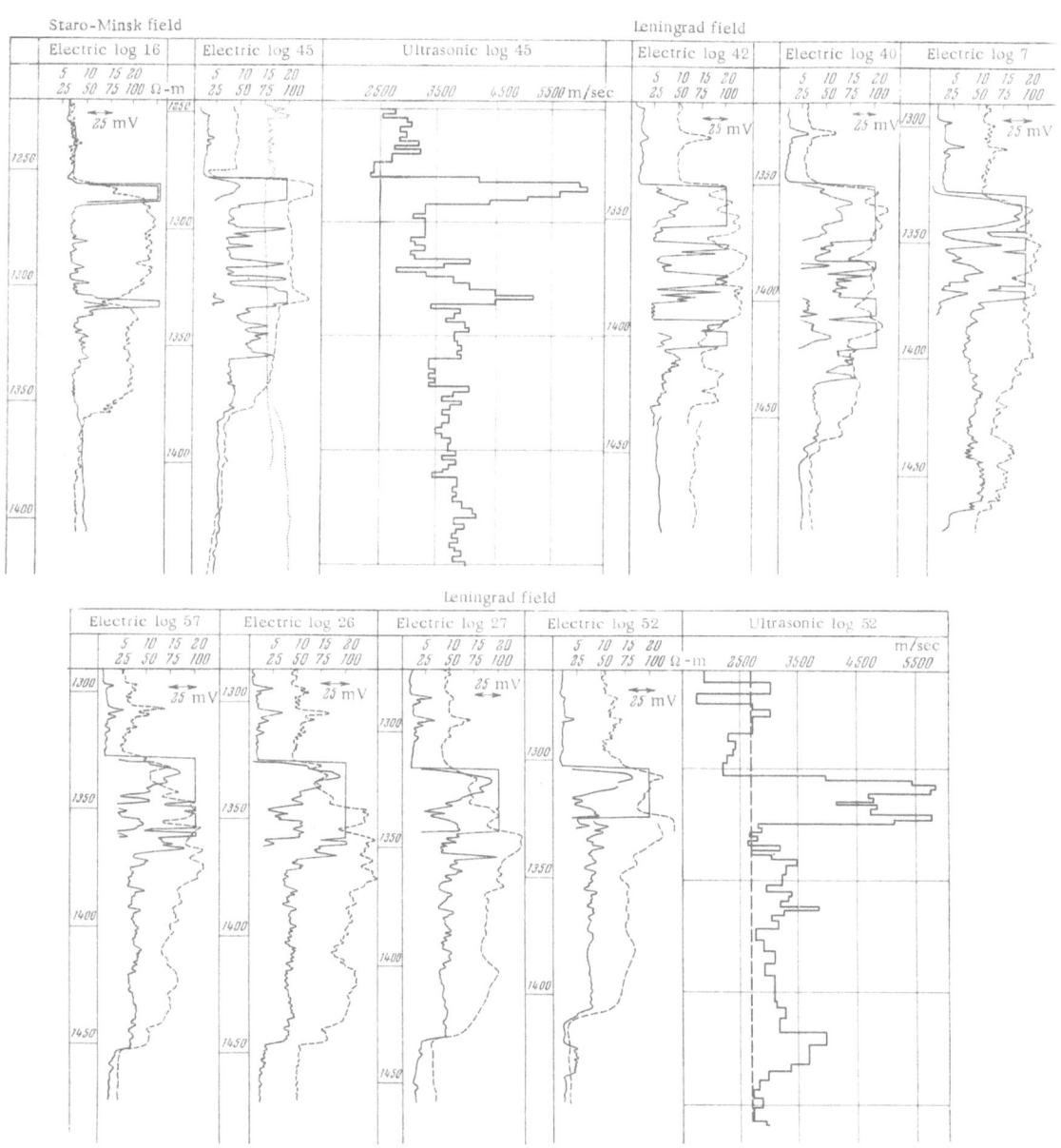

Fig. 6. Nature of bedding in the high-speed layer of the Paleocene sediments
from the Staro-Minsk and Leningrad fields.

TABLE 2

Field	Age — system or subdivision	Age — formation, series or horizon	Well No.	Depth to top, m	Thickness, m	Ultrasonic log data — range in wave speed	Ultrasonic log data — average wave speed	Notes
		Crystalline rocks						
Kushchev	Precambrian		32	1360	Unknown	3.5-5.5	4.4	Transitional layer at top (weathered zone)
		Limestones						
Kushchev	Upper Cretaceous	Santonian	32	1070	110	2.3-3.3	3.1	Thin homogeneous bed with interbeds of lower wave speed
Kushchev	The same		28	1160	130	3.0-3.8	3.3	
Leningrad	"	Tironian	52	1840	20	4.0-4.2	4.3	Interbeds of shale and sandstone
The same	"	Santonian and Kon'yakian	52	1770	20	4.4-6.0	5.1	
" "	Paleogene	Lower Paleocene	52	1300	30	4.0-5.9	5.2	Uniform layer with sharp boundaries
Staro-Minsk	The same		45	1280	11-12	4.6-6.9	5.1	
Staro-Minsk	The same		45	1330	8	4.9-5.5	5.4	
		Marls						
Kushchev	Upper Cretaceous	Kampanian	32	870	8-9	2.7-3.0	2.9	
Leningrad	Lower Cretaceous	Albian	52	2030	145	2.9-4.2	3.4	
	Upper Cretaceous	Senomian	52	1860	40	3.6-4.4	3.8	
		Sandstones and siltstone						
Kushchev	Lower Cretaceous	Albian	32	1300	18-20	2.55-2.85	2.6	Sandstone with siltstone
The same	Upper Cretaceous	Kampanian	32	1000	10	2.35-2.5	2.4	
	Paleogene	Middle zóne	32	830	15	2.1-2.65	2.35	
	"	Oligocene Maikopian	32	400	12	1.7-2.0	1.8	
Leningrad	"	Lower Paleocene	52	1300	60	2.8-3.2	3.1	Laminated with shale
The same	"	The same	39	1380	70	2.6-3.5	3.1	
"	"	Eocene	39	1040	20	2.04-2.2	2.04	
"	"	"	39	1080	25	2.0-2.2	2.10	
"	"	"	39	1155	17	2.2-2.4	2.3	
"	Neogene	Miocene	52	370	18	1.9-2.2	2.05	
"	"	"	IV	370	15	1.7-2.5	2.4	
"	"	Upper Pontian	52	280	20	1.9-2.7	2.7	
"	"	"	IV	275	12	4.7-2.9	2.5	Repernian seismic horizon

TABLE 2 (Continued)

Field	Age: system or subdivision	Age: formation series or horizon	Well No.	Depth to top, m	Thickness, m	Ultrasonic log data: range in wave speed	Ultrasonic log data: average wave speed	Notes
Sands								
Krilov			16	1530	45	1.8-3.2	3.0	With interbeds of siltstone
The same			16	1390	18	2.6-3.2	2.9	
" "			16	1348	17	3.0-3.4	3.2	
" "			16	1285	18	2.6-3.2	2.8	
" "			–	1000	24	1.9-2.7	2.5	
" "			–	730	15	1.9-2.4	2.2	
Leningrad	Neogene	Miocene Kimmeriisian	52	210	26	1.85-2.2	1.9	
Shales								
Kushchev	Upper Cretaceous	Kampanian	32	900	140	2.3-2.6	2.4	
The same	Paleogene	Upper Eocene	32	540	110	1.9-2.2	2.15	
" "	"	Oligocene, Maikopian	32	325	45	1.75-1.95	1.8	
Leningrad	Lower Cretaceous	Albian	52	1910	90	3.0-3.3	3.2	Vertical gradient in wave speed
The same	Upper Cretaceous	Kampanian	52	1450	120	2.3-3.1	2.95	
" "	Paleogene	Upper Eocene	52	1050	130	2.1-2.6	2.25	
" "	"	Oligocene, Maikopian	52	750	90	1.9-2.8	2.4	
" "	Neogene	Miocene, Pontian	52	300	60	1.7-2.0	1.8	
" "	"	Miocene, Kul'nitsian	52	30	100	1.6-2.1	1.8	
" "	Upper Cretaceous	Kampanian	39	1480	80	2.6-3.5	3.1	
" "	The same	Upper Paleocene	39	1260	150	1.8-3.1	2.5	
" "	Paleogene	Upper Eocene	39	870	100	1.9-2.6	2.2	
" "	Neogene	Miocene, Kul'nitsian	IV	30	100	2.4-3.4	2.75	
Krilov			16	1130	150	2.4-2.7	2.5	
The same			16	790	105	1.7-2.2	1.9	
Staro-Minsk	Upper Cretaceous	Kampanian	46	1370	Large 100	2.6-4.1	3.6	Inhomogeneous layer

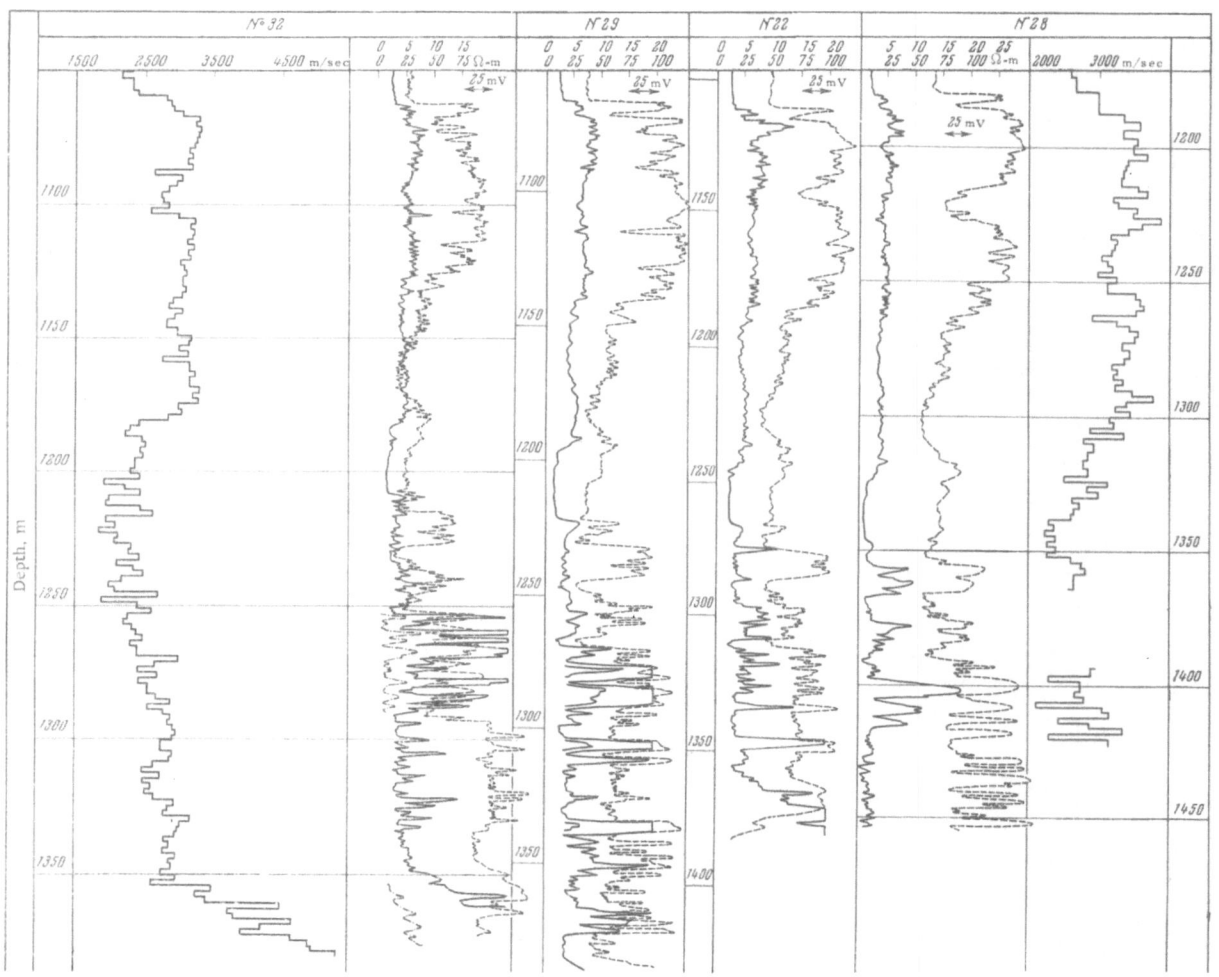

Fig. 7. Correlation of high-speed beds of upper Cretaceous age in the Kushchev field.

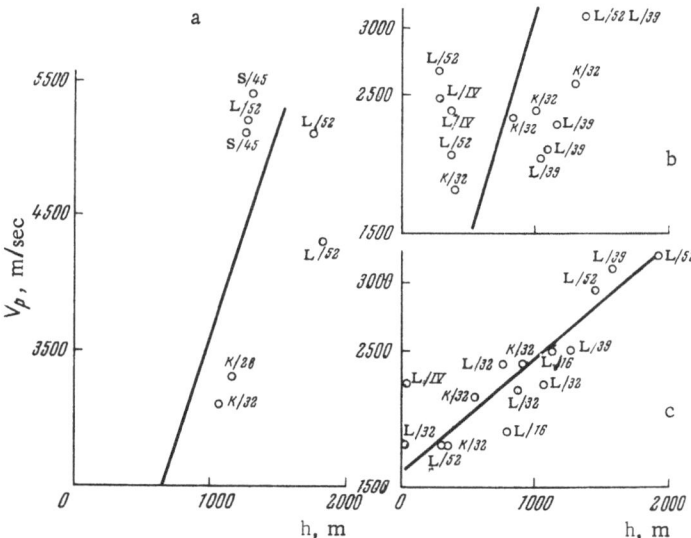

Fig. 8. Wave speeds for various types of rocks. a) Limestones; b) sandstones and siltstones; c) shales. The letters by each point indicate the field (L = Leningrad, K = Kushchev, S = Staro-Minsk), and the number indicates the well.

Consideration of Table 2, along with the graphs shown in Fig. 8, leads to the following generalizations about wave speeds in various types of rocks.

Crystalline rocks were penetrated only to depths within their weathered crust. The wave speed in this part of the section varies from 2.7 to 5.5 km/sec.

Limestones, which are the most consolidated rocks in the upper part of the sedimentary sections penetrated in the northern part of the Krasnodar Belt, are distinguished by a high value for V_P in comparison with other rocks. Paleocene limestones in the Leningrad and Staro-Minsk fields have the highest wave speeds. The wave speed in limestones of Santonian age in the Kushchev field is less than in the limestone of Santonian age in the Leningrad field. This is apparently a result of differences in lithologic character and in depth of burial.

Marls are unimportant in the section. They occur infrequently as thin beds, up to 10 m thick. The wave speed ranges from 2.9 to 3.8 km/sec.

Sandstones are found in nearly all of the sections at depths of 300-1400 m. Most of the individual sandstone beds are only 20-30 m thick. The longitudinal wave speed increases with depth (at 370 m, $V_P = 1.9$ km/sec, while at 1380 m, $V_P = 2.6$ km/sec). On the average, sandstones penetrated at the Leningrad and Staro-Minsk fields have somewhat larger wave speeds than those in the Kushchev field.

The shales are important in the upper part of the section, and individual units have thicknesses up to 150 m. The ultrasonic logs show only minor variations in wave speed (see for example, the 100-240-m interval in Fig. 1, and the 950-1050-m interval in Fig. 4). A clearly apparent increase in wave speed with depth is found for shale (Fig. 8). The vertical gradient in wave speed for shale has an average value of 0.85-0.9 km/sec/km. The shale has a lower wave speed on the average than the other rocks ($\approx$1.7-3.0 km/sec). However, the Kampanian beds in the Staro-Minsk field have a wave speed of about 3.6 km/sec.

Conclusions

Consideration of the longitudinal wave speeds in the upper part of the section in the Northern part of the Krasnodar Belt, based on precise ultrasonic logging, leads to the following conclusions.

1. The general character of the velocity profile was determined (to a depth of 1600-2100 m) and differences were found between the Leningrad, Kushchev, Staro-Minsk, and Krilov fields; the upper part of the section was studied in detail (30-350 m).

2. The average longitudinal wave speed increases with depth; the maximum differences in average wave speeds as determined with sonic and ultrasonic methods are no more than 5%, usually being 1-2%.

3. The following structural features were noted: a) the presence of thin laminated sequences with varying degrees of heterogeneity at various depths in the various fields; b) the existence of sequences of rock in which wave speed increases or decreases with depth (transitional layers); c) thin layers or sequences with high wave speeds which may appear as seismic boundaries forming high-amplitude waves; and d) various types of boundaries between layers (sharp or transitional, strong or weak). These structural features cannot be recognized with low-frequency seismic borehole surveys or with vertical profiling.

4. Some of the characteristics of the velocity profile are persistent over some area, so that they may be taken as constant in seismic exploration.

5. A direct relationship between wave speed and electrical resistivity was used. An increase in V_P is accompanied by an increase in ρ_k. This means that electric logs may be used to provide a recognizable velocity profile in an area. However, further work is required for a quantitative relationship between electric logs and acoustic logs.

6. The average longitudinal wave speeds were determined for the various rock types which comprise the section. Rocks may be listed in the following sequence according to wave speeds: surface zone of the crystalline basement (4.4 km/sec), limestone (3.1-5.4 km/sec), marl (2.9-3.8 km/sec), sandstone (1.8-3.1 km/sec), sand (1.9-3.2 km/sec), and shale (1.8-3.2 km/sec). The wave speed depends on depth in the case of sandstones and shales. The rate of increase with depth is 0.6 to 1.2 km/sec/km.

LITERATURE CITED

1. L. S. Polak and M. B. Rapoport, On the relation between electrical and elastic properties of sedimentary rocks, Prikl. Geofiz., No. 15 (1956).
2. G. I. Petkevich, Factors Determining the Velocity of Seismic Waves in the Geological Section (with the Example of the Fore-Carpathians), Izd. Akad. Nauk UkSSR, 1963.
3. L. Y. Faust, Seismic velocity as a function of depth and geologic time, Geophysics, 16(2) (1951).
4. L. Y. Faust, A velocity function including lithologic variation, Geophysics, 18(2) (1953).
5. Yu. A Bekov, Construction of synthetic seismograms on the basis of neutron-gamma logs or electric logs, Uch. Zap. Permsk. Inst., No. 102 (1963).
6. A. Dziewonski, Obliczanie krzywej predkosci w odwiercie na podstawie danych sondowania elektrycznego, Acta Geophys. Polon., 12(1) (1964).

PART TWO

METHODS FOR AND THE PRINCIPAL RESULTS FROM THEORETICAL STUDIES OF WAVE PROPAGATION

ANALYSIS OF THE SPECTRAL CHARACTERISTICS OF
A THIN-BEDDED SEQUENCE

I.S. Berzon

As a result of the development of spectral methods for interpreting seismic data in recent years [1-6], many methods for computing the spectral characteristics for various representations of layered media have been developed for waves of different types. Analyses have also been made of the properties of the spectral characteristics computed for several representations of the medium. Spectral characteristics have been established for the case of the wave incident on a planar layer consisting of sinusoids. Primarily, the methods which have been developed are valid for determining the spectral characteristic for normal incidence and longitudinal reflection of waves [7-12]. In some of the previous work, methods have been presented for computing the spectral characteristics of a layered medium for the case of inclined wave incidence either using the principle of geometric optics [12, 13] or expressions for wave propagation in a continuous, ponderable medium [14-16]. The method described in references [14, 16] allows the determination of the spectral characteristic for reflected and transmitted waves: longitudinal, volume, and transverse. Here, we will consider only the case of normal incidence of waves on a layer or sequence of layers.

The spectral characteristics for a layer or sequence of layers for reflection of waves at normal incidence have been obtained in analytical form for homogeneous layers [7, 8, 13], for several types of layers with transitional boundaries [9, 12, 13], and for a piecewise transitional layer with boundaries having first-order discontinuities [12].

With the more complex representations of the medium, it is not possible to obtain an exact solution for the spectral characteristic in closed form, and methods have been developed for obtaining the spectral characteristics of sequences of layers using a high-speed computer [10, 11], which allows consideration of multiple reflected waves. In reference [10], the velocity profile of the medium is taken to be that for a combination of transitional layers; all have the same density. In reference [11], the medium is represented as a series of layers separated by first-order boundaries in acoustic impedance. If transitional layers are present in the section, they may be represented with a sequence of beds having first-order boundaries; that this is permissible has been shown in references [17, 18].

In addition to these quite general methods for obtaining spectral characteristics using a computer, there are also approximate methods for obtaining spectral characteristics. With some simplifying assumptions which restrict the generality, these methods allow one to find the spectral characteristics of a layered medium with an analog device [19] or by finding a reasonably simple analytic expression which does not require the use of a computer [20, 21]. This paper deals with the development of an approximate method for computing the spectral characteristics of a layered medium, directed toward obtaining an expression which may be used to evaluate the

47

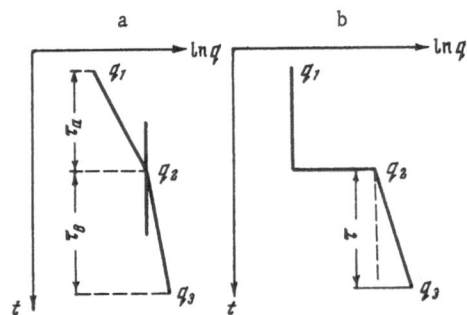

Fig. 1. Schematic representation of boundaries for two transitional layers (a) and for a layer in which there is a discontinuity both in the acoustic impedance and the gradient of acoustic impedance (b).

fundamental behavior of wave reflection. We have considered sequences of layers with first- and second-order boundaries, and in distinction to previous work [11, 19, and others], transitional layers will not be replaced by a sequence of homogeneous layers. We have analyzed the expressions for the amplitude and phase spectrum for several realistic representations, and have made an approximate evaluation of the amplitudes of reflected waves for them.

We have also considered several questions related to the solution of the inverse problem for thin-bedded media, and in particular, the problem about the feasibility of approximating a sequence of layers with a specified type of homogeneous layer with equivalent properties.

§1. Method for the Approximate Computation of the Spectral Characteristics for a Sequence of Layers

The method we used for the approximate computation of the spectral characteristics of a sequence of layers for wave reflection was based on the following assumptions and restrictions.

1. The boundaries between beds within a sequence may be represented as first- or second-order boundaries in acoustic impedance. In any given sequence, both types of boundary may be present, with some boundaries being discontinuities in acoustic impedance and other boundaries being discontinuities in the gradient of acoustic impedance.

2. We have considered only those second-order boundaries in which, for each of the contacting beds, the logarithm of the acoustic impedance q varies linearly with change in the two-way transit time t. In other words, the reflection coefficient R [9, 11, 19]

$$R = \frac{1}{2} \frac{d \ln q}{dt}$$

in each of the two beds is a constant. In this case, the acoustic impedance is an exponential function of t; that is,

$$q = q_0 e^{2Rt}. \tag{2}$$

A special case of Eq. (1) which has been widely used in published papers is the linear relation between propagation speed and depth in a transitional layer with a constant density ρ. In this case, $R = \frac{1}{4}$ grad V (where grad V = dV/dH, and H is depth).

If the manner of variation of acoustic impedance in a bed cannot be presented by Eq. (2), a thin bed may be replaced by a series of very thin layers with a linear law for the variation of the logarithm of acoustic impedance in each of them. This approach is analogous to that of replacing a transitional layer by a series of thinner layers with constant properties [11, 17, 18].

3. We have studied only primary reflection from layer boundaries; refractions along boundaries are not considered. In this case, each boundary within a sequence is treated as an independent reflecting boundary, with the same wave form incident on each. The resulting interference pattern is a consequence of the superposition of primary reflected waves from each of

the boundaries within the sequence. This approximation for waves reflected from a thin layer or sequence of layers has been used previously in references [5, 11, and by others].

Boundaries of First Order. In the case of a first-order boundary, it is well known that the reflected wave has the same form as the incident wave, but the amplitude and polarity of the reflection are determined by the relative acoustic impedances at the boundary.

Boundaries of Second Order. Waves reflected from a second-order boundary have a form which is the integral of the incident wave form $f(t)$ [9, 22]. The amplitude of waves reflected from the boundary of a transitional layer of the type under consideration with a medium having constant acoustic impedance is [9]

$$A = \frac{1}{2} \frac{\ln(q_2/q_1)}{\tau} |f^{(-1)}(t)|_{\max},$$ (3)

where $f^{(-1)}(t)$ is the integral of the incident wave form with respect to time; $|f^{(-1)}(t)|_{\max}$ is the maximum value for the modulus of this function, and τ is the transient time between the points where the values for acoustic impedance are q_1 and q_2. The multiplier $R = \frac{1}{2} \frac{\ln(q_2/q_2)}{\tau}$ characterizes the amplitude of the wave being considered. The polarity of the reflected wave is the same as that of the incident wave form when the gradient in acoustic impedance at the boundary is positive, and the opposite when the gradient is negative.

The boundary between two transitional layers with different gradients in acoustic impedance may be represented as a combination of two boundaries between transitional layers and media with constant acoustic impedance (Fig. 1a). Using Eq. (3) and the correct polarization for waves reflected from a transitional boundary as indicated above, it is possible to write an expression for the amplitude R of the wave reflected from the boundary between two transitional layers as

$$R = -\frac{1}{2} \ln \frac{\ln(q_2/q_1)}{\tau_a} + \frac{1}{2} \frac{\ln(q_3/q_2)}{\tau_b}.$$ (4)

In the special case where grad V_a and grad V_b are of the same size but opposite in sign, the reflection coefficient for the boundary is doubled.

Boundaries at Which There Are Discontinuities in Acoustic Impedance and in the Gradient of Acoustic Impedance. Boundaries of this type may be represented as the combination of two boundaries, one of first order and one of second order (see Fig. 1b). The reflection from such a boundary may also be represented by the superposition of two waves, reflected from boundaries of first and second order. Thus, the total reflected wave is

$$f_{\text{refl}}(t) = k_1 f(t) + \frac{1}{2} \frac{\ln(q_3/q_1)}{\tau} f^{(-1)}(t).$$ (5)

Expanding $\ln q_3/q_2$ in a series in terms of the form $(q_3 - q_2)/(q_3 + q_2)$, the expression for R may be written approximately as

$$R = \frac{q_3 - q_2}{q_3 + q_2} \frac{1}{\tau};$$ (6)

and so, Eq. (5) takes the form

$$f_{\text{refl}}(t) = k_1 f(t) + \frac{k_2}{\tau} f^{(-1)}(t),$$

where

$$k_2 = \frac{q_3 - q_2}{q_3 + q_2}. \tag{5'}$$

Multiple-Bedded Sequences. If the sequence of beds under consideration has n boundaries, of which p are boundaries with discontinuities in acoustic impedance and r are boundaries with a change in the gradient of acoustic impedance, the total reflected wave may be represented in the form

$$f_{\Sigma} = \sum_p k_p f(t - \tau_p) + \sum_r \frac{k_r}{\tau_{lr}} f^{(-1)}(t - \tau_r), \tag{7}$$

where τ_p and τ_r are the time delays for the waves reflected from the various boundaries relative to the wave reflected from the first boundary in the sequence, and τ_{lr} are the transit times in the transitional layers.

The complex spectrum for the reflected waves is

$$\bar{S}_{\Sigma}(\omega) = \bar{S}(\omega) \sum_p k_p e^{-j\omega\tau_p} + \frac{\bar{S}(\omega)}{j\omega} \sum_r \frac{k_r e^{-j\omega_r}}{\tau_{lr}}, \tag{8}$$

where $\bar{S}(\omega)$ is the complex spectrum for the incident wave $f(t)$. The complex reflection coefficient for a sequence of layers or the complex spectral characteristic of the sequence for wave reflection may be written as the ratio of the complex spectrum of the reflected wave to the complex spectrum of the incident wave:

$$\bar{k}(\omega) = \frac{\bar{S}_{\Sigma}(\omega)}{\bar{S}(\omega)} = \sum_p k_p e^{-j\omega\tau_p} + \frac{1}{j\omega} \sum_r \frac{k_q e^{-j\omega\tau_r}}{\tau_{lr}}. \tag{9}$$

Using the expression for the complex reflection coefficient, one may determine the modulus and phase for the reflection coefficient or the amplitude and phase of the frequency characteristic for the layers. Some evaluation of the precision in the case of media with first-order boundaries has been given in references [5, 8, 11, and others]. These computations indicated that for a single layer, the error in determining the reflection coefficient exceeds 10% only for marked contrasts in the speeds of propagation between layers (with $q_1/q_2 < 0.5$). In the case of a periodic sequence of layers with first-order boundaries, the error will not exceed 10% for boundaries at which the contrast in acoustic impedance lies in the range $0.7 < q_2/q_1 < 1.4$. The question of the errors involved in determining the spectral characteristics for a case in which second-order boundaries are present using this approximate method will be considered in the following paragraphs.

§2. Relationship Between the Spectral Characteristics for Beds with Boundaries of First and Second Order

It has been pointed out in §1 that the properties of waves reflected from the boundaries of transitional layers permit the establishment of an analogy between layers or sequences of layers having first-order boundaries and those having second-order boundaries. A second-

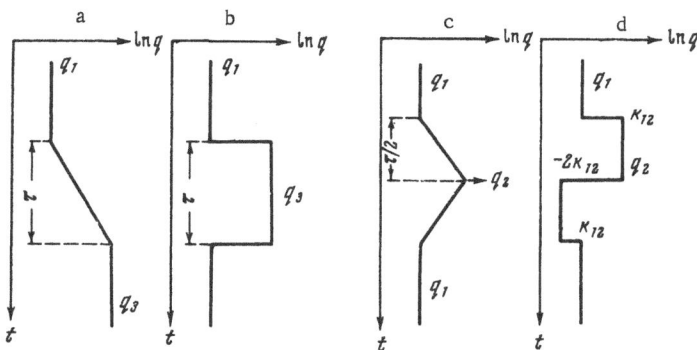

Fig. 2. Representations of transitional layers (a, c) and
their analogs in media with discrete boundaries (b, d).

order boundary at which the velocity gradient increases reflects a wave with the same polarity
as that from a first-order boundary at which the discontinuity in acoustic impedance is positive. We will attempt to establish a quantitative relationship between the spectral characteristics for layers with second-order boundaries and layers with first-order boundaries having
similar values for τ in the layers and similar values for the ratio of acoustic impedances q.
In so doing, we will consider two simple cases: a transitional layer and a double transitional
layer.

Transitional Layer

In media with first-order boundaries, the analogy to a transitional layer with a positive
velocity gradient is a homogeneous layer with increased acoustic impedance (see Fig. 2a, b).
We will consider the reflection coefficient for the transitional layer and then compare it with
the reflection coefficient for a homogeneous layer. According to Eq. (9), the reflection coefficient for a transitional layer is

$$k(\omega) = \frac{k_{13}}{j\omega\tau}(1 - e^{-j\omega\tau}),\qquad(10)$$

where τ is the two-way travel time in the transitional layer and k_{13} is the reflection coefficient for the boundary between two halfspaces with the properties of layers 1 and 3. The
modulus and phase of the reflection coefficient are:

$$k(\omega) = \frac{k_{13}}{\omega\tau/2}\left|\sin\frac{\omega\tau}{2}\right|,\qquad(11)$$

$$\left.\begin{array}{l} \sin\varphi(\omega) = \pm\left|\sin\dfrac{\omega\tau}{2}\right|, \\[2mm] \cos\varphi(\omega) = \pm\cos\dfrac{\omega\tau}{2}\,\dfrac{\sin\dfrac{\omega\tau}{2}}{\left|\sin\dfrac{\omega\tau}{2}\right|}. \end{array}\right\}\qquad(12)$$

The plus and minus signs in Eq. (12) refer to the cases $k_{13} > 0$ and $k_{13} < 0$, respectively. It is
obvious from Eq. (11) that as $\omega\tau \to 0$, $k(\omega) \to k_{13}$, and for $\omega\tau/2 = m\pi$ (where m is an integer),
$k(\omega) = 0$.

With $k_{13} > 0$, the phase $\varphi(\omega)$ increases monotonically from zero to π, as $\omega\tau/2$ increases,
with a discontinuity at π.

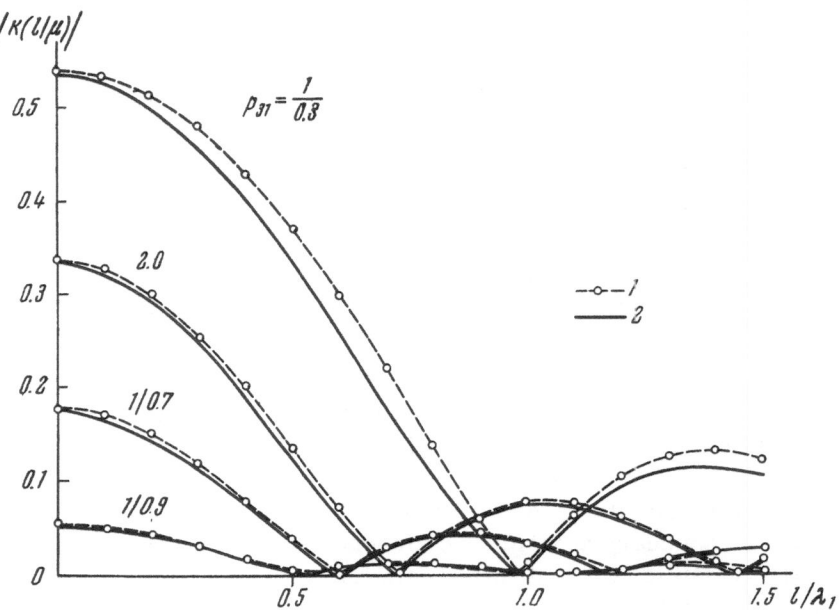

Fig. 3. Amplitude spectrums for a transitional layer with different values for the ratio p_{31} with respect to the surrounding halfspaces. 1) Exact; 2) approximate.

Comparison with Exact Computations.   We will now compare the results that we have obtained with the exact solution for a transitional layer with a linear change in speed of propagation as a function of depth, with constant density [14]. We will write the expression for the two-way travel time through the layer in the form

$$\tau = \frac{2l}{V_3 - V_1} \ln \frac{V_3}{V_1},$$

so that Eqs. (11) and (12) take the forms

$$k(l/\lambda_1) = \frac{(p_{31} - 1)^2}{(p_{31} + 1)\ln p_{31}} \frac{1}{2\pi l/\lambda_1} \sin\left(\frac{2\pi l/\lambda_1 \ln p_{31}}{p_{31} - 1}\right),$$

(11')

$$\varphi = \frac{2\pi \ln p_{31}}{p_{31} - 1} l/\lambda_1,$$

(12')

where $p_{31} = V_3/V_1$, $\lambda_1 = V_1/f$ is the wave length of the incident wave and f is the frequency in cps.

Figure 3 shows curves for $k = k(l/\lambda_1)$ calculated both from Eq. (11) and from the exact expressions [13, 2]. The curves for $\varphi(l/\lambda_1)$ obtained with the approximate solution are the same as those obtained with the exact expression [12]. It is apparent from Fig. 3 that only for large values for p_{31}, ≥ 3.33, does the difference in values for $k(l/\lambda_1)$ amount to 15% for any value of l/λ_1. Results which have been obtained for small differences in the magnitude of the reflection factor evaluated for a transitional layer using the exact and the approximate expressions, as well as the coincidence in phase for the reflection factor, agree well with computations of synthetic seismograms, either with or without the consideration of multiples [22]. These computations have indicated that the approaches are equivalent for practical purposes.

On the basis of these results, we may assume that the approximate method for computing spectral characteristics may also be used for a sequence of layers in which there are second-order boundaries.

Comparison with Reflection Factors for a Homogeneous Layer with Increased or Decreased Acoustic Impedance. For the case of a homogeneous layer with a reflection factor k_{13} at the boundary, the complex reflection factor computed with the approximate method is

$$\bar{k}(\omega) = k_{31}(1 - e^{-j\omega\tau}).\tag{14}$$

It follows from Eq. (14) that the magnitude and phase of the reflection factor for a homogeneous layer are

$$k(\omega) = 2\left| k_{13}\sin\frac{\omega\tau}{2}\right|,\tag{15}$$

$$\left.\begin{aligned}\cos\varphi &= \pm\left|\sin\frac{\omega\tau}{2}\right|,\\\sin\varphi &= \pm\frac{\sin\frac{\omega\tau}{2}}{\left|\sin\frac{\omega\tau}{2}\right|}\cos\frac{\omega\tau}{2}.\end{aligned}\right\}\tag{16}$$

It follows from this example that the complex reflection factor for a transitional layer differs from that for a uniform layer with a contrast in acoustic impedance only by the multiplying term $1/j\omega\tau$. Considering the properties of multiplication of complex quantities, one needs only to multiply the magnitude of the reflection factor for a uniform layer by $1/\omega\tau$, and increase its phase by $\pi/2$ to obtain the magnitude and phase of the reflection factor for the transitional layer. On the basis of these simple considerations, it follows that in the case of a transitional layer, the heights of the secondary lobes in the spectrum are inversely proportional to frequency, while the phase discontinuity at π is shifted to $\varphi = 0°$. As $\omega \to 0$, the terms $1/\omega\tau$ and $k(\omega)$ for a homogeneous layer both increase without limit, but their product has a finite limit k_{13}.

Symmetrical Double Transitional Layer

An example of layers of this type in a medium with only first-order boundaries would be that of two homogeneous layers having the same two-way travel times, with reflection factors of the same size but opposite in sign (see Fig. 2c, d). We will consider the reflection factor for transitional layers and compare it with the reflection factor for a two-layer sequence with first-order boundaries.

Let the total transit time in the layers be τ, so the transit time for each layer individually is $\tau/2$. The spectrum for the reflection factor may be written in the form

$$\bar{k}(\omega) = \frac{k_{12}}{\tau/2}\frac{1}{j\omega}\left(1 - e^{\frac{j\omega\tau}{2}}\right)^2,\tag{17}$$

where k_{12} is the reflection factor computed from the values $q_1 = \rho_1 V_1$ in the medium outside the layers and $q_2 = \rho_2 V_2$ at the inflection point on the acoustic impedance profile (see Fig. 2c).

Writing out the magnitude and phase for the complex reflection factor, we have

$$\bar{k}(\omega) = 4k_{12}\frac{\sin^2\frac{\omega\tau}{4}}{\frac{\omega\tau}{2}}\left(\sin\frac{\omega\tau}{2} + j\cos\frac{\omega\tau}{2}\right);\tag{18}$$

$$k(\omega) = \frac{4\,|\,k_{12}\,|\,\sin^2\frac{\omega\tau}{4}}{\omega\tau/2}\;;\tag{19}$$

$$\varphi(\omega) = \pm\left(-\frac{\pi}{2} + \frac{\omega\tau}{2}\right).\tag{20}$$

It is apparent from Eq. (19) that $k(\omega) = 0$ when $\omega\tau/4 = n\pi$ (where n is integer) or when $f\tau = 2n$. The function $k(\omega)$ has minimums at these values for $f\tau$ (see Fig. 4a). The phase of the reflection factor increases linearly from $-\pi/2$ at $f\tau = 0$ (see Fig. 4b).

Comparison with the Reflection Factor for a Double Homogeneous Layer. For a double homogeneous layer, the magnitude and phase of the reflection factor are

$$k(\omega) = 4\,|\,k_{12}\,|\,\sin^2\frac{\omega\tau}{4}\;,\tag{21}$$

$$\varphi(\omega) = \pm\left(\pi + \frac{\omega\tau}{2}\right).\tag{22}$$

Curves are shown in Fig. 4 for $k(f)/k_{12}$ and $\varphi(f)$ for $k_{12} > 0$.

It is apparent from these examples that the following method might be used for computing the reflection factors for simple types of layers with second-order boundaries: 1) comparison with the analogous case with first-order boundaries and computation of the magnitude and phase of the reflection factor for this; and 2) application of a correction for the transitional boundaries to these results by multiplying the magnitude of the reflection factor by $1/\omega\tau_u$ (where τ_u is the transit time in a basic layer) and increasing the phase by $\pi/2$.

§3. Periodic Sequences of Layers

In this section, we consider three types of periodic sequences, with examples shown in Fig. 5.

Periodic Sequence Consisting of Uniform Layers with Constant Acoustic Impedances

and Uniform Transit Times in Each Layer (Fig. 4a)

In our examination of this problem, we will assume that the acoustic impedance in the medium outside the sequence is the same as that in the layers in the sequence with the lower

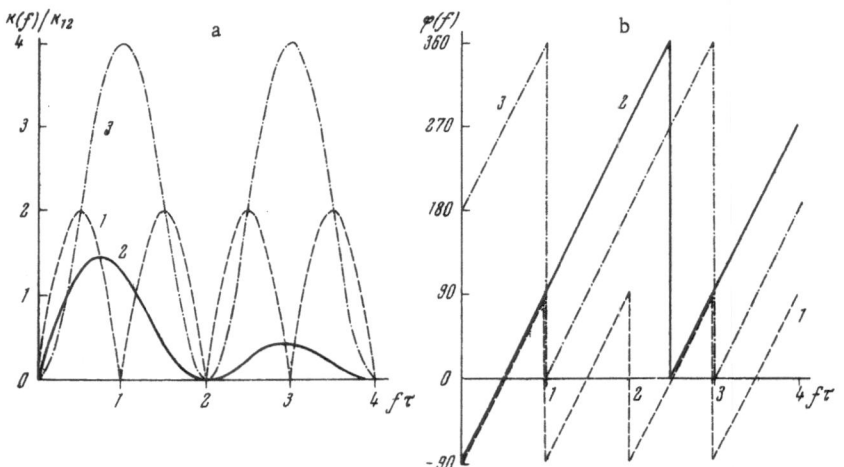

Fig. 4. Amplitude (a) and phase (b) spectrums. 1) Homogeneous layer; 2) double transitional layer; 3) double homogeneous layer.

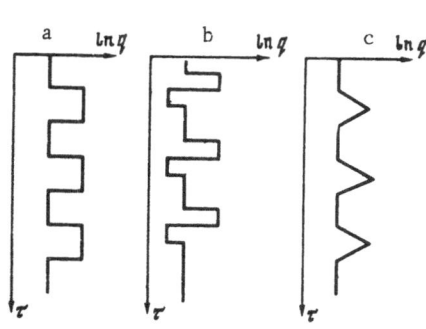

Fig. 5. Representations of periodic sequences under consideration.

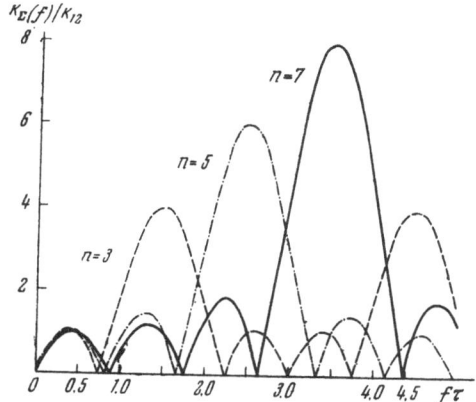

Fig. 6. Amplitude spectrum for the sequence defined in Fig. 5a.

values of acoustic impedance. Using Eq. (9), we can write the expression for the complex reflection factor as a series

$$\bar{k}_\Sigma(\omega) = k_{12}(1 - e^{-j\omega\tau_1} + e^{-2j\omega\tau_1} - e^{-3j\omega\tau_1} + \cdots - e^{-nj\omega\tau_1}) \qquad (23)$$

or

$$\bar{k}_\Sigma(\omega) = k_{12}(1 - e^{-j\omega\tau_1})(1 + e^{-2j\omega\tau_1} + e^{-4j\omega\tau_1} + \cdots + e^{-(n-1)j\omega\tau_1}), \qquad (24)$$

where τ_1 is the two-way travel time in a single layer. Using trigonometric representations and considering that

$$\sin 2\omega\tau_1 + \sin 4\omega\tau_1 + \cdots + \sin(n-1)\omega\tau_1 = \frac{\sin\dfrac{n+1}{2}\omega\tau_1 \sin\dfrac{n-1}{2}\omega\tau_1}{\sin\omega\tau_1}, \qquad (25)$$

$$1 + \cos 2\omega\tau_1 + \cos 4\omega\tau_1 + \cdots + \cos(n-1)\omega\tau_1 = 1 + \frac{\cos\dfrac{n+1}{2}\omega\tau_1 \sin\dfrac{n-1}{2}\omega\tau_1}{\sin\omega\tau_1}, \qquad (26)$$

it is possible to express the magnitude of the reflection factor as

$$k_\Sigma(\omega) = 2\left|k\sin\frac{\omega\tau_1}{2}\right|\sqrt{1 + 2\frac{\cos\dfrac{n+1}{2}\omega\tau_1 \sin\dfrac{n-1}{2}\omega\tau_1}{\sin\omega\tau_1} + \frac{\sin^2\dfrac{n-1}{2}\omega\tau_1}{\sin^2\omega\tau_1}}. \qquad (27)$$

If we use the symbol

$$\tau = n\tau_1,$$

to represent the travel time through the whole sequence, Eq. (27) may be written as

$$k_\Sigma(\omega) = 2\left|k_{12}\sin\frac{\omega\tau}{2n}\right|\sqrt{1 + 2\frac{\cos\dfrac{n+1}{2}\dfrac{\omega\tau}{n} \sin\dfrac{n-1}{2}\dfrac{\omega\tau}{n}}{\sin\dfrac{\omega\tau}{n}} + \frac{\sin^2\dfrac{n-1}{2}\dfrac{\omega\tau}{n}}{\sin^2\dfrac{\omega\tau}{n}}}. \qquad (28)$$

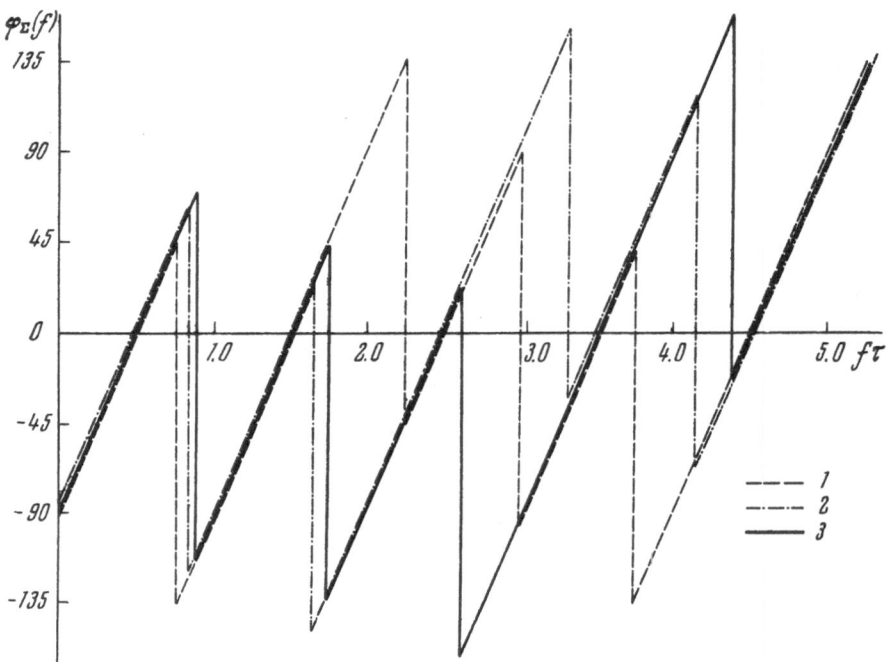

Fig. 7. Phase spectrum for the sequence defined in Fig. 5a. 1) n = 3;
2) n = 5; 3) n = 7.

<u>Funadmental Properties of the Magnitude of the Reflection Factor</u>.
The product of two terms comprises Eq. (24) and (27). The first of these is the approximate
expression for the relationship between the reflection factor and frequency for a single layer
with sharp boundaries [see Eq. (15)]. The second term, as is apparent from Eq. (24), is analogous to the directional characteristic for a detector pattern consisting of $p = (n + 1)/2$ equally
sensitive elements; the difference in arrival times between two neighboring elements is $2\tau_1$
[23]. It is obvious that the second term is a periodic function of $f\tau = \omega\tau/2\pi$ with a fundamental period of $n/2$. The maximum values for this term are found at $f\tau = 0$ and $f\tau = n/2$, and the
zero values are found at

$$f\tau = s\frac{n}{n+1} \quad \left(\text{where } s = 1, 2, \ldots, \frac{n-1}{2} \right).$$

The first term in the product in Eq. (28) — the function $|k_{12} \sin(\omega\tau/2n)|$ — has zeros at
$f\tau = 0$ and $f\tau = n$. Thus, the product of these two terms is also periodic, with a fundamental
period $f\tau = n$, with a principal maximum at $f\tau = n/2$ with zeros distributed at intervals $\Delta f\tau$
$= n/(n+1)$, except about the principal maximums, where $\Delta f\tau = 2n/(n+1)$, and with subsidiary
maximums located between each consecutive pair of zeros (see Fig. 6). The maximum value
is $k_{\Sigma}(f\tau) = k_{12}(n+1)$. It is apparent that only values for $k_{\Sigma}(f\tau) < 1$, have physical significance,
and so, the following inequality may be used:

$$k_{12}(n+1) < 1. \tag{29}$$

With the use of this inequality, evaluations of $k_{\Sigma}(f\tau)$ contain errors as a consequence of
the approximations used.

<u>Phase of the Reflection Factor</u>   The phases of waves reflected from a se-
quence of layers, measured relative to the phases of the waves incident at the top of the se-

quence, may be given as

$$\sin \varphi = \pm \frac{\sin \frac{\omega\tau}{n}\cos \frac{\omega\tau}{2n} + \sin \frac{n-1}{2n}\omega\tau \cos \frac{n+2}{2n}\omega\tau}{\sin \frac{\omega\tau}{n}\sqrt{1+2\frac{\cos \frac{n+1}{2}\frac{\omega\tau}{n}\sin \frac{n-1}{2}\frac{\omega\tau}{n}}{\sin \frac{\omega\tau}{n}} + \frac{\sin^2 \frac{n-1}{2}\frac{\omega\tau}{n}}{\sin^2 \frac{\omega\tau}{n}}}} \frac{\sin \frac{\omega\tau}{2n}}{\left|\sin \frac{\omega\tau}{2n}\right|}; \qquad (30)$$

$$\cos \varphi = \pm \frac{\sin \frac{\omega\tau}{2n}\sin \frac{\omega\tau}{n} + \sin \frac{n-1}{2n}\omega\tau \sin \frac{n+2}{2n}\omega\tau}{\sin \frac{\omega\tau}{n}\sqrt{1+2\frac{\cos \frac{n+1}{2n}\omega\tau \sin \frac{n-1}{2n}\omega\tau}{\sin \frac{\omega\tau}{n}} + \frac{\sin^2 \frac{n-1}{2}\frac{\omega\tau}{n}}{\sin^2 \frac{\omega\tau}{n}}}} \frac{\sin \frac{\omega\tau}{2n}}{\left|\sin \frac{\omega\tau}{2n}\right|}; \qquad (31)$$

with the upper sign corresponding to the case $k_{12} > 0$, and the lower sign to the case $k_{12} < 0$. The phase of the reflection factor tends to $-\pi/2$ for $k_{12} > 0$ and $f\tau = 0$, as it did for a homogeneous layer. At the points where the magnitude of the reflection factor has zeros, the phase has a discontinuity with an amplitude of π (see Fig. 7). The phase passes through zero at $f\tau = 0.5$, 1.5, and so on.

The periodicity of the phase spectrum, as well as that of the amplitude spectrum, is $f\tau = n$, or $\tau_1 = 1$; that is, it is determined by the transit time through a single layer τ_1.

The basic differences between the phase spectrums for a sequence of layers and a single layer are:

1) The zeros of the phase spectrum do not coincide with the subsidiary maximums of the amplitude spectrum; and

2) at points where the phase jumps by π, the value does not shift from $\pi/2$ to $-\pi/2$, as in the case of a single layer, but between other values which depend on the number of layers in the sequence and on the location of a particular zero value in the amplitude spectrum.

Sequences Consisting of Double Homogeneous and Double Transitional Layers (Fig. 5b, c)

The complex spectrum for the sequence shown in Fig. 5b is

$$\overline{k}_{\Sigma}(\omega) = k_{12}(1 - 2e^{-j\omega\tau_1/2} + e^{-j\omega\tau_1})(1 + e^{-2j\omega\tau_1} + e^{-4j\omega\tau_1} + \cdots + e^{-(n-1)j\omega\tau_1}). \qquad (32)$$

As was the case earlier, the spectrum is the product of two terms. One of these is the complex spectrum for a single double-layer while the other is the spectrum for a discrete distribution of layers, corresponding to the characteristic of a pattern of geophones or shot holes.

For the sequence of transitional layers defined in Fig. 5c, it is possible to develop the spectrum using relationships between the spectrums for transitional layers and for layers with sharp boundaries (see §2). The spectrum for a sequence of transitional layers differs from Eq. (32) only by the presence of the multiplier $1/j\omega\tau_1/2$. The magnitude of the reflection factor for a sequence of transitional layers is

$$k_{\Sigma}(\omega) = \frac{4|k_{12}|\sin^2 \frac{\omega\tau}{4n}}{\frac{\omega\tau}{2n}}\sqrt{1+2\frac{\cos \frac{n+1}{2}\frac{\omega\tau}{n}\sin \frac{n-1}{2}\frac{\omega\tau}{n}}{\sin \frac{\omega\tau}{n}} + \frac{\sin^2 \frac{n-1}{2}\frac{\omega\tau}{n}}{\sin^2 \frac{\omega\tau}{n}}}, \qquad (33)$$

where τ is the two-way travel time through the whole sequence.

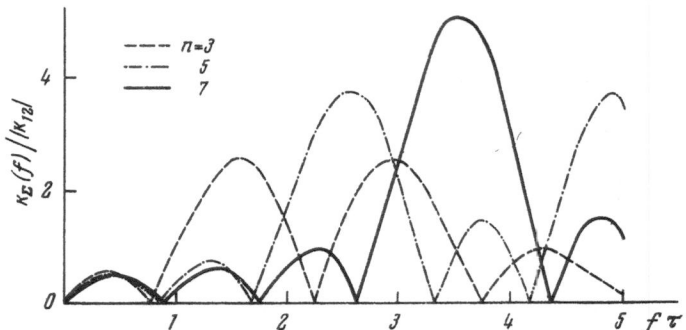

Fig. 8. Amplitude spectrum for the sequence defined in Fig. 5c.

In view of the fact that the first term, which is the spectrum for a double transitional layer, has zeros at intervals $f\tau$, which are twice as large as in the case of a uniform layer with sharp boundaries, the curve for $k_{\Sigma}(f)$ (Fig. 8) will not be symmetrical about $f\tau = n/2$, as in the case of Fig. 6. The positions of those zeros in the spectrum $k_{\Sigma}(f)$ which are determined by the radical remain the same as in Fig. 6, but the ratios of amplitudes of the maximum points are quite different.

The phase spectrum for the transitional layers differs from the phase spectrum for the case shown in Fig. 7 only in the absence of the multiplying term $\sin \omega\tau/2n$, so that for $f\tau = n$, the phase spectrum for a double transitional layer does not have a jump of π, as it does in the case of a uniform layer with sharp boundaries (Fig. 4b). Thus, with $f\tau \leq n$, the phase spectrums for both types of sequence coincide, but with $f\tau > n$, they differ by an amount π.

From a consideration of the material presented in this section, it is evident that the amplitude and phase spectrums for a periodic sequence may be essentially different than those for homogeneous layers.

§4. Approximate Determination of the Amplitudes of Waves

Reflected from a Thin-Bedded Sequence

Basic Formulas. In determining the amplitudes of waves reflected from a thin-bedded sequence, we will make use of Rayleigh's theorem for the energy in a pulse [24]:

$$E = \frac{1}{\pi} \int_0^{\infty} S^2(\omega)\, d\omega = \int_{-\infty}^{\infty} f^2(t)\, dt, \tag{34}$$

where $S(\omega)$ is the magnitude of the amplitude spectrum of the waves, and $f(t)$ is the pulse being considered. The energy content of the waves $E^{(t)}$, evaluated in the time domain, may be written as [25]

$$E^{(t)} = \int_{-\infty}^{+\infty} f^2(t)\, dt \cong \bar{a}^2 (t_2 - t_1), \tag{35}$$

where $\bar{a}$ is the average value for the amplitude spectrum, and $t_2 - t_1$ is the duration of the pulse. This is equivalent to replacing the actual record with segments of a sinusoid with amplitude $\bar{a}$.

The duration of a wave is dependent on the range of frequencies contained in its spectrum [24]:

$$t_2 - t_1 = \frac{c}{f_l - f_k},$$ (36)

where f_l and f_k are the frequency limits of the spectrum and c is a parameter with a value which varies slightly for waves of different forms [24]. As indicated in [24], the particular frequency limits taken affect the results to some extent; here, we will take as frequency limits two consecutive frequencies at which the amplitude spectrum has zeros.

The amplitude spectrum for reflected waves, for which the energy is being considered, may be written in the form

$$S(\omega) = Q(\omega) k_\Sigma(\omega),$$ (37)

where $Q(\omega)$ is the spectrum of the incident waves and $k_\Sigma(\omega)$ is the magnitude of the reflection factor for a layer or sequence of layers.

In this section, we consider the possible amplitudes for reflected waves for two representations of the medium (Figs. 2d and 5a), and compare these amplitudes with the amplitudes of waves reflected from the surface of a halfspace. First, we will consider a much simplified case in which, over the frequency range from zero to some cutoff frequency f_g, the spectrum for the incident wave is a horizontal straight line; then we will consider examples in which the spectrum of the incident wave is more band limited.

Double Layer with First-Order Boundaries (Fig. 2d)

Wideband Spectrum for the Incident Waves. Consider that the spectrum of reflected waves is given over a range of values for $f_g \tau$ from zero to 2.0, corresponding to a single period in the spectrum for a double layer. On the basis of the assumption concerning the spectrum of the incident wave, the wide-band energy $E_{wb}^{(f)}$ for the reflected waves, expressed as a spectrum, is

$$E_{wb}^{(f)} = \frac{16}{\pi} k_{12}^2 Q^2 \int_0^{\omega_g} \sin^4 \frac{\omega\tau}{4} d\omega = 12 k_{12}^2 Q^2 f_g.$$ (38)

Using Eqs. (35) and (36), we find that the energy $E_{wb}^{(f)}$ is

$$E_{wb}^{(f)} = \overline{a}^2 \frac{c_1}{f_g},$$ (39)

where $C_1 = \sqrt{C}$.

It follows from Eqs. (38) and (39) that

$$\overline{a}_{wb} = \frac{2\sqrt{3}}{c_1} k_{12} Q f_g$$ (40)

We now compare the expressions which have been obtained for the energy and average amplitude with the corresponding cases of a halfspace boundary and a homogeneous layer having the same thickness as a single member of the double layer. In this case, the zeros for the spectrums for the homogeneous layer and the double layer coincide.

Comparison with Waves Reflected from the Surface of a Halfspace.
The energy $E_h^{(f)}$ of a wave reflected from the surface of a halfspace, expressed as a spectrum,
is

$$E_h^{(f)} = 2k_{12}^2 Q^2 f_g, \tag{41}$$

while the energy expressed in the time domain is given by Eq. (39). Thus, the amount of energy
reflected from a double layer is larger by a factor of six than the energy reflected from the
surface of a halfspace. This means that the ratio of amplitudes for waves reflected from a
double layer and from a halfspace is

$$\frac{\bar{a}_{wb}}{\bar{a}_h} = \sqrt{6} \approx 2.5.$$

Comparison with Waves Reflected from a Uniform Layer. The energy $E_c^{(f)}$ for a wave reflected from a uniform layer is

$$E_c^{(f)} = \frac{4k_{12}^2 Q^2}{\pi} \int_0^{\omega_g} \sin^2 \frac{\omega\tau}{2} d\omega = 4k_{12}^2 Q^2 f_g, \tag{42}$$

that is, it is twice as large as the energy reflected from a halfspace. The ratio of average
amplitudes in the case of a double and of a uniform layer is

$$\frac{\bar{a}_{wb}}{\bar{a}_c} \cong \sqrt{3},$$

and that for waves reflected from a uniform layer and a halfspace boundary is

$$\frac{\bar{a}_c}{\bar{a}_h} \cong \sqrt{2}.$$

Narrow-Band Spectrum for the Incident Waves. In a real medium, the
spectrum of the incident wave is rarely as broad as that considered in the preceding paragraphs. We will now assume that $Q(f)$ is a constant over some range of frequencies from f_1
to f_2 and zero outside this range. We will evaluate the energy in waves reflected from a double
layer, a halfspace, and a uniform layer, as well as the ratios of amplitudes of these waves, under this assumption.

For the double layer

$$E_{wb}^{(f)} = \frac{4k_{12}^2 Q^2}{\pi}\left[\frac{3}{2}\pi(f_2 - f_1) - \frac{4}{\tau}\left(\sin\frac{\pi f_2\tau}{2} - \sin\frac{\pi f_1\tau}{2}\right) + \frac{1}{2\tau}(\sin 2\pi f_2\tau - \sin 2\pi f_1\tau)\right]; \tag{43}$$

for the halfspace

$$E_h^{(f)} = 2k_{12}^2 Q^2 (f_2 - f_1); \tag{44}$$

and for the uniform layer

$$E_c^{(f)} = \frac{2k_{12}^2 Q^2}{\pi}\left[2\pi(f_2 - f_1) - \frac{1}{\tau}(\sin 2\pi f_2\tau - \sin 2\pi f_1\tau)\right]. \tag{45}$$

We will evaluate the amplitudes of the reflected waves for the case in which the spectrum of the incident waves coincides with the region of maximum amplitude in the spectrums
for the double and uniform layers. In this case, the amplitude of waves reflected from a layer

(either double or uniform) must be a maximum. Taking $f_1 = f_g/4$ and $f_2 = 3f_g/4$, we have

$$E_{wb}^{(f)} = 11.1 k_{12}^2 Q^2 f_g, \tag{43'}$$

$$E_h^{(f)} = k_{12}^2 Q^2 f_g, \tag{44'}$$

$$E_c^{(f)} = 3.26 k_{12}^2 Q^2 f_g. \tag{45'}$$

Using Eq. (39) for $E^{(f)}$, we can write the ratios of average amplitudes for these various reflected waves

$$\overline{a}_{wb} : \overline{a}_c : \overline{a}_h = 3.3 : 1.8 : 1.0.$$

These ratios indicate that for thin layers (uniform or double), the maximum possible amplitude for reflected waves may be considerably larger than the amplitude of a wave reflected from a halfspace. We might note that even if a reflected coefficient of $2k_{12}$, which is that for the boundary at the middle of the double layer, is taken for the surface of a halfspace, the average amplitude of the wave reflected from the double layer is still larger by a factor of 1.65 than the amplitude of a wave reflected from the halfspace.

Periodic Sequences with Well-Defined Boundaries (Fig. 5a)

We will now evaluate the amplitudes of waves reflected from a thin layered sequence for two cases:

1) The transit time τ_1 in each layer of the sequence is the same, but the number of layers is variable, as is the total thickness of the sequence; and

2) the total transit time τ through the sequence is invariant, but the number of layers is variable, as is the transit time τ_1, in each of the layers.

Sequences with Variable Thickness (Case 1). The periodicity f_m of the spectrum remains fixed with respect to the frequency axis as the number of layers is increased. The spacing Δf between adjacent zeros in the spectrum varies according to the relation

$$\Delta f = \frac{f_m}{n+1}, \tag{46}$$

with Δf about the principal maximum being twice as wide as the others.

Broad-Band Spectrum for the Incident Wave. We will examine the simplified case in which the spectrum of the incident wave is $Q(f) = $ const over the range from 0 to f_m.

For a periodic layered sequence with n layers, the spectrum $k_\Sigma(\omega)$, and consequently the spectrum $S(\omega)$, will contain n lobes over the frequency range under consideration. The energy $E^{(f)}$ may be represented in the form of a sum of individual energy terms, each corresponding to a lobe in the spectrum $S(\omega)$:

$$E(\omega) = \frac{k_{12}^2 Q^2}{\pi} \left[\int_0^{\omega_1} S_1^2(\omega)\, d\omega + \int_{\omega_1}^{\omega_2} S_2^2(\omega)\, d\omega + \cdots \right]. \tag{47}$$

The area of each lobe may be obtained approximately as the area of a triangle with a base equal to $\Delta\omega = 2\pi\Delta f$ for the subsidiary maxima and $2\Delta\omega$ for the principal maximum, and with a height S_k^2 between $Q^2 k_{12}^2$ and $2Q^2 k_{12}^2$ for the subsidiary maxima and equaling $Q^2 k_{12}^2 (n+1)^2$ for the principal maximum. Studies have shown that the total energy contained in the subsidiary maxima does not amount to more than a few percent of the energy contained in the principal maximum. Therefore, we may ignore the subsidiary maxima in an approximate evaluation. Then, the energy in a reflected wave may be written approximately as

$$E^{(f)} \approx \frac{k_{12}^2 Q^2}{\pi} \int\limits_{\frac{\omega_m}{2} - \Delta\omega}^{\frac{\omega_m}{2} + \Delta\omega} S_k^2(\omega)\, d\omega \approx \frac{k_{12}^2 Q^2 (n+1)^2}{2\pi} \cdot \frac{2\omega_m}{n+1} = 2k_{12}^2 Q^2 (n+1) f_m. \tag{48}$$

Thus, it follows that for a fixed f_m the energy in a wave increases with the number n of layers.

In evaluating the average amplitudes in Eqs. (35) and (36), it is necessary that the duration of the oscillations be expressed over the range of frequencies $f_l - f_k$ in the spectrum. According to an earlier derivation,

$$f_l - f_k = \frac{2f_m}{n+1} \tag{46'}$$

and so,

$$E^{(f)} \approx \overline{a}^2 \, \frac{c(n+1)}{2f_m}. \tag{49}$$

As may be seen from Eqs. (36) and (46), the duration of a wave oscillation increases in proportion to $n+1$ as the number n of layers is increased. Therefore, as follows from Eqs. (48) and (49), the average amplitude $\overline{a}$ of a wave does not depend on the number of layers, that is,

$$\overline{a}_n \approx \overline{a}_p \approx \text{const.} \tag{50}$$

Thus, in the type of sequence under consideration, consisting of beds with constant τ, the amplitude of a reflected wave is invariant with respect to the number of layers. In the case under consideration with identical layers comprising the sequence, we use the condition $f\tau_1 = 0.5$ or $l_1/\lambda = 0.25$.

As has been indicated earlier, for the same type of spectrum for the incident wave, the average amplitude of the wave reflected from a thin layer is larger by $\sqrt{2}$ than the average amplitude of the wave reflected from the surface of a halfspace. Exactly the same amplitude relation holds for sequences consisting of layers with the same transit time τ_1.

Narrow-Band Spectrum for the Incident Wave. Let the spectrum of the incident wave be given over a narrower band of frequencies, corresponding to the band of frequencies in the principal maximum for $n = 3$. Then, the result developed earlier about the essential constancy of the amplitude with an increasing number of layers still holds for $n \geq 3$. However, the relationship of amplitudes for waves reflected from a sequence consisting of three layers and for waves reflected from a single layer is changed, as may be seen from Eqs. (45'), (48), and (49). The ratios of average amplitudes for waves reflected from a three-layer sequence, a single layer, and a halfspace are

$$\overline{a}_{n=3} : \overline{a}_{n=1} : \overline{a}_h = 2.9 : 1.8 : 1.0.$$

Sequence with a Fixed Total Transit Time and a Variable Number of Layers (Case 2). With a fixed total transit time τ in a sequence and an increase in the number of layers, with a corresponding decrease in the transit time in each, the frequency limit for the fundamental period in the spectrum increases, and the principal maximum shifts toward higher frequencies.

Substituting n/τ in place of f_m in Eq. (48), we have

$$E^{(f)} = \frac{2k_{12}^2 Q^2 (n+1)}{\tau}.$$

Therefore, the energy of a reflected wave increases with increasing n. Thus, the expression for the ratio of energies for n and k layers in a sequence is

$$\frac{E_n^{(f)}}{E_k^{(f)}} = \frac{n(n+1)}{k(k+1)}. \tag{51}$$

The spacing Δf between zeros about the principal maximum is

$$\Delta f = \frac{2n}{(n+1)\tau}. \tag{52}$$

Using Eq. (52), we may write the duration of the oscillations in a reflected wave as

$$\Delta t = \frac{c\tau(n+1)}{2n}. \tag{53}$$

Thus, the duration of the wave train even diminishes slightly with increasing n. We obtain an expression for the ratio of energies for the cases of n or k layers from Eqs. (35) and (53):

$$\frac{E_n^{(t)}}{E_k^{(t)}} = \frac{\bar{a}_n^2 (n+1)k}{\bar{a}_k^2 (k+1)n}. \tag{54}$$

Equating the right-hand parts of Eqs. (51) and (54), we obtain

$$\frac{\bar{a}_n}{\bar{a}_k} = \frac{n}{k}. \tag{55}$$

Consequently, with an increasing number of layers in a sequence with a given total transit time τ, the average amplitude of the reflected waves may increase if, naturally, the incident wave contains spectral components about the principal maximum of the spectrum for the sequence. Thus, for example, if the two-way travel time in a sequence is $\tau = 0.06$ sec and the spectrum of the incident wave is flat from 12 to 62 cps, the average amplitude increases by a factor of 1.6 in changing the number of layers from 3 to 5. Similarly, the average amplitude will be increased by a factor of 1.4 if the incident spectrum is flat over the narrower spectrum from 30–70 cps and the number of layers is increased from 5 to 7. With a further increase in the number of layers, the maximum in the spectrum will shift to still higher frequencies, which usually will not be as strong in the incident wave, and therefore the average amplitudes of waves reflected from the sequence must diminish.

§5.   Some Aspects of the Solution to the Inverse Problem

On the Possible Detailed Resolution of Actual Sequences

The question of the possible detailed resolution of a medium on the basis of the spectrum for longitudinally reflected waves for uniform layers has been considered in references [1-5], as well as others. More complicated cases in which the sequence consists of layers with variable transit time have been considered in a paper by Kats in this collection. Here, we will consider only cases of periodic sequences consisting of layers with first-order boundaries with the same transit time in each layer (see §3). Such an approach permits an evaluation of the least thickness for a layer which may be identified from its spectral characteristics in media with differing properties.

It follows from §3 that the periodicity of the spectrum is

$$f\tau = n$$

or

$$f\tau_1 = 1,$$

where $\tau_1 = \tau/n$ is the transit time in a single layer. Hence, it follows that for a limiting frequency f_g, the spectrum for the minimum transit time in a single layer which may be recognized from its spectrum is

$$\tau_1 = \frac{1}{f_g}. \tag{56}$$

For a limiting frequency $f_g = 100$ cps, τ_1 is 0.01 sec. The thickness of the layer is related to the time τ_1 by the expression

$$\frac{2l_i}{V_i} = \tau_1.$$

Consequently,

$$l_i = \frac{V_i}{2f_g}.$$

For a reasonable range in propagation speeds in an actual medium from 1500 to 6000 m/sec, l_1 is 7.5 to 30 m when $f_g = 100$ cps. These numbers represent realistic limits for the discrimination of a sequence from its spectrum. These figures are significantly better than the layer thicknesses which may be determined by direct measurements of propagation speed in a medium using ultrasonic logging. This should be kept in mind in the solution of the inverse problem — the construction of a model of the medium according to wave propagation and the comparison of such a derived model with ultrasonic logging data.

In order to obtain a significant increase in resolution of a sequence of layers it would be necessary to go to higher frequencies; an increase in the cutoff frequency by a factor k would allow the resolution of layers thinner by the factor k for the same propagation speed. A significant contribution to the solution of this problem might apparently be the development of a shooting pattern which would allow the selective reinforcement of the high-frequency components of seismic waves.

In view of the fact that detailed resolution of a thin-bedded sequence is not practically possible in many cases, sometimes a sequence is considered to be a quasiuniform bed with constant effective properties. We will consider the question of some possible limitations in determining the effective properties of a sequence in layers.

Determining the Effective Properties of a Sequence of Layers

At present, methods for determining the effective properties of a sequence of layers from the amplitude spectrum are used in cases where a quasiperiodic distribution of minimum points is noted along the frequency axis in the spectrum. At the same time, the following differences between observed spectrums and those calculated from theory for a uniform layer may be noted: 1) differences in the amplitudes of the various maximums; and 2) the maximums are sharper and the minimums smoothed out in the observed spectrum. These differences are caused by the departure of the structure of an actual medium from that assumed for the mathematical representation. We will consider the possible use of methods for determining the effective properties, calculated for the representation of uniform layer, for other representations which have spectrums qualitatively the same as that for a uniform layer. We will consider two simple examples:

1) A double layer with first-order boundaries; and

2) a periodic sequence with equal transit times τ_1 in each layer.

Double Layer with First-Order Boundaries. If we take the layers as uniform, the effective transit time t_f given by the frequency separation between two consecutive minimums is

$$\tau_f = \frac{\tau}{2},$$

where τ is the true transit time for the double layer. If the average speed of propagation in the layers is known accurately, then the layer thickness computed from the spectral characteristics will be too small by a factor of two.

We will evaluate the error in determining the thickness of a layer which is possible in cases in which the average propagation speed in the sequence is determined from the amplitude spectrum [2, 4].

According to Eq. (15), the maximum value for the reflection factor from a uniform thin layer is $2k_{12}$.* Taking the spectrum for a two-layer sequence as that for a homogeneous layer, we obtain an expression for the effective ratio of acoustic impedances between the medium about the layer and the layer itself.

$$q_{12}^f = \frac{1 - 2k_{12}}{1 + 2k_{12}}. \tag{57}$$

If we assume that the densities are the same in both media, then after some algebraic manipulation, it is possible to obtain the following relationship between the effective and true average speeds of propagation in a double layer from Eq. (57):

$$\frac{\overline{V}_{2f}}{\overline{V}_2} = \frac{(3 - q_{12})(1 + q_{12}^2)}{2q_{12}(3q_{12} - 1)}. \tag{58}$$

---

*With consideration of multiple reflections in the layer, this value will be somewhat smaller and tends to $2k_{12}$ as $\rho_1 V_1 / \rho_2 V_2 \to 1$. However, even for relatively strong contrasts in acoustic impedance ($q_{12} = 0.5$) this difference does not exceed 10%, and we will ignore it.

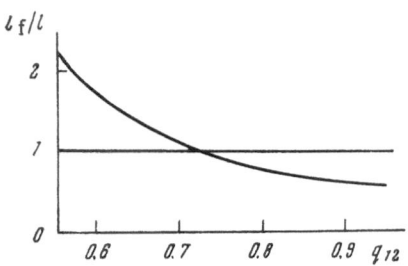

Fig. 9. Curve for the relationship between l_f/l and the ratio of acoustic impedances q_{12} obtained by approximating the amplitude spectrum for a double layer with the spectrum of a uniform layer.

This corresponds to the ratio

$$\frac{l_f}{l} = \frac{1}{4} \frac{(3 - q_{12})(1 + q_{12}^2)}{q_{12}(3q_{12} - 1)} \cdot \tag{59}$$

The relationship between l_f/l and q_{12} is shown in Fig. 9. It is apparent from this curve that for small q_{12}, the effective thickness of the layer may be much overestimated because the effective average propagation speed determined from the amplitude spectrum is larger than the true average speed. For large q_{12} that is, for small differences in speeds, the value for $\overline{V}_f/\overline{V}_2$ tends to unity and the thickness may be underestimated.

From this example, it follows that prior to using the method for determining effective properties for an equivalent uniform thin layer, the validity of the approach should be checked. In this case, as well as in a number of others, the phase spectrum may be used as a criterion, inasmuch as the spectrum for a sequence lacks a discontinuity of π which is present in the spectrum for a homogeneous layer (Fig. 4b).

Periodic Sequence (Fig. 5a). If the part of the amplitude spectrum located within the range of the principal maximum approximates the spectrum for a uniform layer, the ratio of the effective τ_f and the true τ is

$$\frac{\tau_f}{\tau} = \frac{n+1}{n} . \tag{60}$$

Hence

$$\frac{l_f}{l_t} = \frac{n+1}{n} \frac{V_f}{V_t} , \tag{61}$$

where l_t and V_t are the corresponding true thickness and average speed in the sequence.

It follows from Eq. (61) that if a velocity V_f equal to the true velocity V_t in the sequence is taken in the interpretation, the error $(l_f - l_t)/l_t = \Delta l/l$ takes the form

$$\frac{\Delta l}{l_t} = \frac{1}{n} . \tag{62}$$

Since the approximation of the spectrum for a sequence of a uniform layer is permissible for $n \geq 5$, the relative error in determining the layer thickness when V_t is known does not exceed 20%. The case in which the speed V_t is unknown beforehand and must be determined along with the thickness from the spectrum has been considered in reference [11]. It has been shown in this paper that the spectrum for a sequence of layers is approximated quite well by the spectrum for a uniform layer with a thickness close to the true value and a propagation speed close to that for low-frequency waves in the sequence. Equation (61) is in agreement with these results inasmuch as the speed of low-frequency waves (in this case V_f) is somewhat less than the speed V_t of high-frequency waves.

The phase characteristic may serve as a criterion indicating the inhomogeneity of a sequence. The value for τ determined from consecutive zeros in the phase spectrum differs from that determined from the amplitude spectrum and equals the true value for τ.

Conclusions

1. We have presented an approximate method for computing the spectral response of a sequence of layers for wave reflection. The layers in the sequence may be separated by first- or second-order boundaries, and some boundaries may be discontinuities in acoustic impedance while other boundaries are discontinuities in the gradient of acoustic impedance.

2. For some representations of layers and sequences which are quite realistic, an analysis has provided approximate expressions for the amplitude and phase spectrums, and characteristic properties by which these representations may be distinguished from one another are noted.

3. Approximate determinations of the amplitudes of waves reflected from thin-layered sequences have been made based on the use of Rayleigh's theorem concerning the energy in a wave form. This study showed that the amplitude of waves reflected from various sequences may in some cases be several times larger than the amplitude of waves reflected from a half-space with the same change in propagation speed.

4. It has been shown that the approximation of the spectrum for a thin-layered sequence with the spectrum for a uniform layer may lead to large errors in determining the effective properties of the sequence in some cases, such as the thickness and the propagation speed for longitudinal waves. The phase spectrum may be used to test the validity of such an approximation.

LITERATURE CITED

1. L. L. Khudzinskii, On determining some properties of layers of intermediate thickness from the spectrum of reflected waves, Izv. Akad. Nauk SSSR, Ser. Geofiz., No. 5 (1961).
2. I. S. Berzon, A. M. Epinat'eva, G. N. Pariiskaya, and S. P. Starodubrovskaya, Dynamic Characteristics of Seismic Waves in Real Media, Izd. Akad. Nauk SSSR, 1962.
3. S. P. Starodubrovskaya and G. N. Pariiskaya, Use of the dynamic characteristics of reflected waves for identifying and tracing of variable thickness, Razvedochnaya Geofiz., No. 2 (1964).
4. I. S. Berzon, On determining models of thin-layered media for the combined use of the amplitude and phase spectrums of layers, Izv. Akad. Nauk SSSR, Ser. Fiz. Zemli, No. 6 (1965).
5. L. L. Khudzinskii, Description of the use of spectral analysis in seismic exploration, in: Deep Seismic Sounding of the Earth's Crust in the USSR, Gostoptekhizdat, 1961.
6. N. E. Grin, On the question of the spectrums of waves reflected from inclined layers, Tr. Inst. Geofiz. Akad. Nauk SSSR, No. 7 (1964).
7. I. I. Gurvich, On reflections from thin layers in seismic exploration, Prikl. Geofiz., No. 9 (1952).
8. I. S. Berzon, On some spectral characteristics of waves reflected from thin layers, Izv. Akad. Nauk SSSR, Ser. Geofiz., No. 5 (1959).
9. J. G. J. Scholte, Propagation of waves in inhomogeneous media, Geophys. Prospecting, 9(1) (1961).
10. L. H. Berryman, P. L. Goupillard, and K. H. Waters, Reflections from multiple transition layers, Geophysics, 24(2) (1959).
11. N. G. Mikhailova, B. S. Pariiskii, and M. V. Saks, Frequency spectrum of a sequence of layers, Izv. Akad. Nauk SSSR, Ser. Fiz. Zemli, No. 1 (1966).
12. O. K. Kondrat'ev, Analytic solution of the problem of reflection from some inhomogeneous layers, Izv. Akad. Nauk SSSR, Ser. Fiz. Zemli, No. 8 (1965).
13. L. M. Brekhovskikh, Waves in Layered Media, Izd. Akad. Nauk SSSR, 1967.
14. G. S. Pod'yapol'skii, Coefficients for refraction and reflection of elastic waves on a layer, Izv. Akad. Nauk SSSR, Ser. Geofiz., No. 4 (1961).

15. J. G. J. Scholte, Oblique propagation of waves in inhomogeneous media, Geophys. J. (of the Roy. Astr. Soc.), 7(2) (1962).

16. L. I. Ratnikova and A. L. Levshin, Calculation of the spectrum of a thin layered medium (oblique incidence), Izv. Akad. Nauk SSSR, Ser. Fiz. Zemli, No. 2 (1967).

17. R. Bortfield, Seismic waves in transition layers, Geophys. Prospecting, Vol. 8 (1960).

18. J. Chauveau, Contribution a l'étude de la deformation du signal seismique Analogies Filtrage equivalent, Geophys. Prospecting, 10(4) (1962).

19. L. L. Khudzinskii, On determining some spectral characteristics of layered media, Izv. Akad. Nauk SSSR, Ser. Geofiz., No. 3 (1962).

20. B. I. Bespyatov, Frequency method for studying the conditions for formation of reflections from thin-layered media, Tr. Nizhnevol. Nauchno-Issled. Inst. Geolog. i Geofiz., Vol. 1 (1964).

21. V. S. Isaev, Results of the application of grouping theory for the study of periodic sequences of layers, Prikl. Geofiz., No. 49 (1967) [English translation: Exploration Geophysics, Vol. 49, Consultants Bureau, New York, 1969].

22. N. G. Mikhailova and B. S. Pariiskii, Computation of synthetic seismograms for simplified cases of structure in the medium with normal incidence, Izv. Akad. Nauk SSSR, Ser. Geofiz., No. 1 (1964).

23. I. I. Gurvich, Seismic Exploration, Gostoptekhizdat, 1960.

24. A. A. Kharkevich, Spectra and Analysis, GITTL, 1957 [English translation: Consultants Bureau, New York, 1960].

25. T. G. Rautian and L. S. Samoilova, On the basis for computing energy density by the method of approximating a seismic trace by segments of sinusoids, Tr. Inst. Fiz. Zemli, "Fizika Zemletryasenii i Seismika Vzryvov," No. 25:192 (1962).

SPECTRAL ANALYSIS FOR RECOGNITION OF WAVE FORMS

S.A. Kats

Originally, spectral analysis was used in seismic exploration to separate two overlapping waves [1-3]. Such methods were based on the use of diagnostic features for the frequency response of a bed. The basis for the method of determining the thickness of a thin bed from diagnostic points on its frequency response characteristic is the assumption that waves are reflected in a similar manner from both the top and the bottom of the bed. This assumption is also made in this presentation. Earlier approaches are generalized in this development by restricting neither the number of interfering waves nor the ratios between their delay times. It is assumed only that the number of interfering waves is finite.

Statement of the Problem

A complicated wave form $\varphi(t)$ made up of several simple wave forms $a_k f(t - \tau_k)$ is recorded:

$$\varphi(t) = \sum_k a_k f(t - \tau_k). \tag{1}$$

We will assume that the form of the simple wave $f(t)$ is known. We must determined the amplitude a_k and the time delay τ_k. Geologically, such a problem might be met, for example, in interpreting wave forms for reflections from a thin-layered sequence, containing beds with a slight dip, with dip angles being of the order 1-5°.

The following spectral representations are well known:

$$A(\omega) = \int_{-\infty}^{\infty} \varphi(t) e^{-i\omega t} \, dt,$$

$$S(\omega) = \int_{-\infty}^{\infty} f(t) e^{-i\omega t} \, dt.$$

In the following interpretation of a wave form reflected from a thin-layered medium, $S(\omega)$ is the spectrum of the incident wave and $A(\omega)$ is the spectrum of the reflected wave.

Applying the Fourier transform to both sides of Eq. (1), we have

$$R(\omega) = \frac{A(\omega)}{S(\omega)} = \sum_k a_k e^{-i\omega\tau_k}. \tag{2}$$

The function $R(\omega)$ will be called the frequency spectrum for a group of waves. When $A(\omega)$ and $S(\omega)$ are known, the spectrum $R(\omega)$ is also known. Thus, the problem of separating a group of waves into its component parts reduces to the problem of determining values for a_k and τ_k from the experimentally determined frequency spectrum for that group.

The method of solving this problem consists of the use of a special transformation of the frequency spectrum $R(\omega)$ to a function $y(v)$, where the argument v may be real or complex. This transformation will be defined in such a way that diagnostic points on the graph of $|y(v)|$ or on the graphs of Real $y(v)$ or Imaginary $y(v)$ may be used to determine values for a_k and τ_k.

We will examine two classes of transformations:

1) transformation of $R(\omega)$ to a function with diagnostic extremals; and

2) transformation of $R(\omega)$ to a function with approximately a step-like behavior.

Determination of the Structure of Interfering Waves from an Integral Transform of a Band-Limited Frequency Spectrum

We will examine several methods for analyzing interfering waves. These methods may be used independently of one another, so that the same result may be obtained by various methods. By so doing, the reliability of the result may be improved.

Transforming $R(\omega)$ into a Function with Diagnostic Extremals

In solving this problem, we will assume that $R(\omega)$ is band limited in the band (ω_1, ω_2).

Fourier Transform of the Frequency Spectrum:

$$S(v) = \int_{\omega_1}^{\omega_2} R(\omega) e^{i\omega v}\, d\omega = 2 \sum_k a_k e^{i\frac{\omega_1 + \omega_2}{2}(\tau_k - v)} \cdot \frac{\sin \frac{\omega_1 - \omega_2}{2}(\tau_k - v)}{\tau_k - v} \cdot \tag{3}$$

$$S(v) = \sum_k a_k S_0 (v - \tau_k).$$

The modulus of the k-th term in the sum (3) will be

$$|S_0(v - \tau_k)| = 2\left| \frac{\sin \frac{\omega_1 - \omega_2}{2}(\tau_k - v)}{\tau_k - v} \right|. \tag{4}$$

The function $|S_0(v - \tau_k)|$ has a principal maximum at $v = \tau_k$, and decreases rapidly as v varies from τ_k; this decrease is stronger with larger values for $|\omega_2 - \omega_1|$. If the values for $(\tau_{k+l} - \tau_k)$ and $(\omega_2 - \omega_1)$ are sufficiently large, then

$$|S_0(v - \tau_{k+l})| \ll |S_0(v - \tau_k)| \tag{5}$$

for $v \simeq \tau_k$.

If Eq. (5) holds, then

$$S(v) \simeq S_0(v - \tau_k),$$

$$v \sim \tau_k.$$

In this case, the curve for the function $|S(v)|$ will have k maxima, located at the points $v = \tau_k$. The amplitude of the k-th maximum will be

$$S(\tau_k) = 2a_k S_0(0) = 2a_k(\omega_2 - \omega_1).$$

We obtain the following expressions for the real and imaginary parts of S(v):

$$\operatorname{Re} S(v) = \sum_k \cos\left[\frac{\omega_1 + \omega_2}{2}(\tau_k - v)\right] \cdot \sin\left[\frac{\omega_1 - \omega_2}{2}(\tau_k - v)\right]\frac{a_k}{\tau_k - v},$$

$$\operatorname{Im} S(v) = \sum_k \sin\left[\frac{\omega_1 + \omega_2}{2}(\tau_k - v)\right] \cdot \sin\left[\frac{\omega_1 - \omega_2}{2}(\tau_k - v)\right]\frac{a_k}{\tau_k - v}.$$

(6)

As may be seen, $|\operatorname{Re} S(v)|$ has maxima at the same values as the function $|S(v)|$, while $|\operatorname{Im} S(v)|$ does not.

Thus, it is possible to determine the time delays and the amplitudes for interfering waves from the maxima on the curve for $|S(v)|$ [3]. By comparing the curves for $|S(v)|$, $|\operatorname{Re} S(v)|$, and $|\operatorname{Im} S(v)|$, it is possible to identify the maxima on the curve for $|S(v)|$ which are associated with real waves. These maxima are duplicated on the curves for $|S(v)|$ and $|\operatorname{Re} S(v)|$, but absent on the curve for $|\operatorname{Im} S(v)|$.

Fourier Transform of the Real and Imaginary Parts of the Frequency Spectrum. We will consider the Fourier transform for the real and imaginary parts of the frequency spectrum over the range $(-\omega_2, \omega_2)$:

$$A_{\operatorname{Re}}(v) = \int_{-\omega_2}^{\omega_2} \operatorname{Re} R(\omega)e^{i\omega v}\,d\omega = 2\sum_k a_k \frac{v\cos(\omega_2\tau_k)\sin(v\omega_2) - \tau_k\sin(\omega_2\tau_k)\cos(v\omega_2)}{\tau_k^2 - v^2};$$

(7)

$$A_{\operatorname{Im}}(v) = \int_{-\omega_2}^{\omega_2} \operatorname{Im} R(\omega)e^{i\omega v}\,d\omega = 2i\sum_k a_k \frac{v\cos(v\omega_2)\sin(\omega_2\tau_k) - \tau_k\sin(v\omega_2)\cos(\omega_2\tau_k)}{\tau_k^2 - v^2}.$$

(8)

Each of the terms comprising $A_{\operatorname{Re}}(v)$ or $A_{\operatorname{Im}}(v)$ is respectively the spectrum of a sinusoid or cosinusoid with a duration of $2\omega_2$ and with a frequency τ_k, which has a maximum at

$$v \approx \tau_k.$$

The amplitudes of $A_{\operatorname{Re}}(v)$ and $A_{\operatorname{Im}}(v)$, for sufficiently large ω_2 and sufficiently large values for $(\tau_{k+l} - \tau_k)$ will be

$$A_{\operatorname{Re}}(v) \simeq a_k\left(\omega_2 + \frac{\sin 2\tau_k\omega_2}{4\tau_k}\right),$$

$$A_{\operatorname{Im}}(v) \simeq a_k\left(\omega_2 - \frac{\sin 2\tau_k\omega_2}{4\tau_k}\right).$$

(9)

In order that both the time delays τ_k and the amplitudes of the interfering waves may be determined from the extremals of the curves for $A_{\operatorname{Re}}(v)$ and $A_{\operatorname{Im}}(v)$, it is necessary to use both transforms, so that

$$a_k = \frac{A_{\operatorname{Re}}(\tau_k) + A_{\operatorname{Im}}(\tau_k)}{2}.$$

(10)

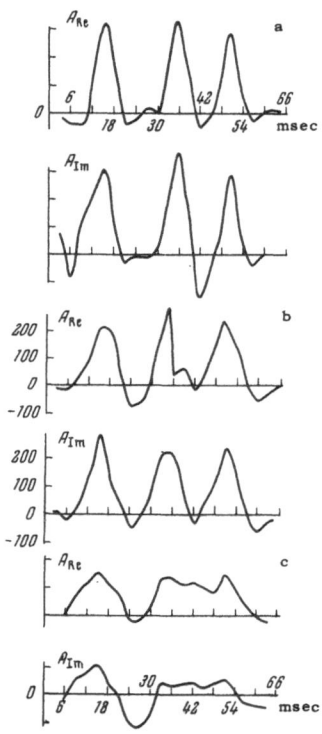

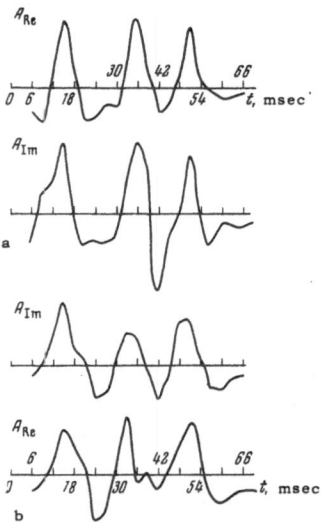

Fig. 1. Curves for the functions $A_{Re}(v)$ and $A_{Im}(v)$ for various high-frequency cutoffs. a) $\omega_2 = 200\pi$ rad/sec; b) $\omega_2 = 140\pi$ rad/sec; c) $\omega_2 = 100\pi$ rad/sec.

Fig. 2. Curves for the functions $A_{Re}(v)$ and $A_{Im}(v)$ for the case of a low-frequency cutoff at 10 cps. a) High-frequency cutoff at 100 cps; b) high-frequency cutoff at 70 cps.

The functions $A_{Re}(v)$ and $A_{Im}(v)$ are shown in Fig. 1 for the case of three interfering impulses:

$$\varphi(t) = \sum_{k=1}^{3} a_k f(t - \tau_k), \tag{11}$$

where $a_k = 1$, $\tau_1 = 0.015$ sec, $\tau_2 = 0.035$ sec, and $\tau_3 = 0.05$ sec.

With these values for τ_k, the impulses $f(t - \tau_k)$ cannot be distinguished from one another, and a record of the combined wave form is a complicated vibration $\varphi(t)$. Curves have been drawn for the functions $A_{Re}(v)$ and $A_{Im}(v)$ for three values of ω_2; 200 π, 140 π, and 100 π rad/sec. As may be seen, with bandwidths of 100 and 70 cps, it is possible to determine the time delays τ_k quite well, while the amplitude factors a_k can be determined with a precision of about 10%. For $\omega_2 = 100\pi$ rad/sec, τ_k cannot be determined so easily, but it may still be done.

The spectrum $R(\omega)$ cannot be determined at frequencies close to zero because seismic equipment cannot record low frequencies. Therefore, different approaches to determining $R(\omega)$ at frequencies $\omega < \omega_L$ have been found to be proper, where ω_L is the low-frequency cutoff of the seismic equipment. The simplest of these is based on the assumption that

$$R(\omega) = 0 \quad \text{for} \quad |\omega| < \omega_L. \tag{12}$$

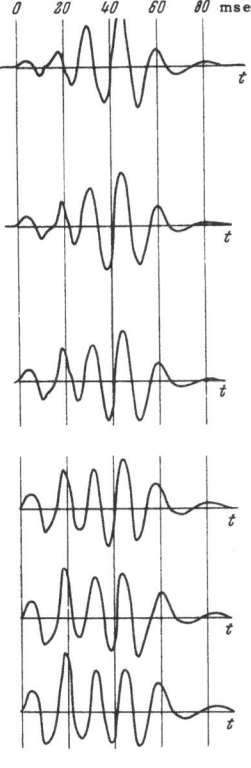

Fig. 3. Synthetic
seismograms for
cases of interfer-
ence between waves
with different am-
plitudes.

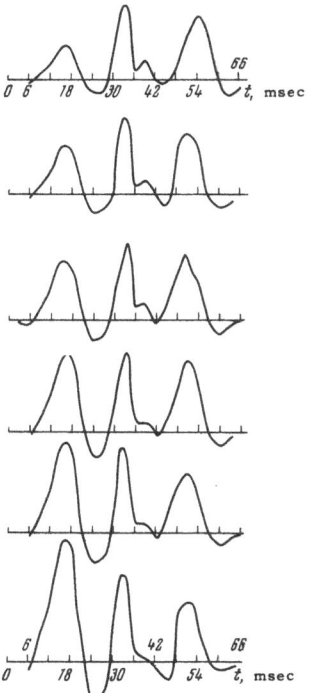

Fig. 4. Curves for
$A_{Re}(v)$ correspond-
ing to the synthetic
seismograms in
Fig. 3.

The effect of a low-frequency cutoff, $F_L = 10$ cps, on the functions $A_{Re}(v)$ and $A_{Im}(v)$, is
shown in Fig. 2. It is evident that the loss of the low-frequency part of the spectrum causes
no difficulty in the determination of the actual time shifts on the record, even though the general picture is more complicated. False extremals are present, with amplitudes the same as
for actually recorded events. In this case, the false extremals differ from the true ones in
sign. One may attempt to distinguish the false extremals by comparing the functions A_{Re} and
A_{Im}. Thus, for example, the extremal for $\tau_k = 0.042$ sec which is apparent on the curve for
$A_{Im}(v)$ in Fig. 2 is only weakly expressed on the curve for $A_{Re}(v)$. In view of this, it may be
discarded.

We have computed functions $A_{Re}(v)$ for interfering waves with differing amplitudes, in
order to evaluate the dynamic range obtainable with the technique. We considered the superposition of three waves, the first of which was assigned an amplitude which was varied from
0.5 to 2. The two later waves both had unit amplitudes. The corresponding synthetic seismograms and spectrums are shown in Fig. 3 and Fig. 4. As may be seen from the spectrums,
for this range in variation of the amplitudes of the component waves, they may be distinguished
from one another quite adequately. If the ratio of amplitudes is made greater than three, the
waves cannot be readily separated, because the amplitudes of the false extremals begin to mask
the amplitudes of the extremals corresponding to the true waves.

Transformation of R(ω) into a Staircase Function

A transformation which converts the frequency spectrum into a function having extremals which can be used to determine values for a_k and τ_k is not the only satisfactory way for analyzing interfering waves. It is possible to find a group of transformations which change the frequency spectrum, or its real or imaginary part to a function which is approximately a staircase. The amplitude of the k-th step will be proportional to a_k, while the distance between the k-th and the (k + 1)th step will be proportional to

$$(\tau_{k+1} - \tau_k).$$

This transformation has the form

$$\varphi_1(v) = \int_{-\omega_2}^{\omega_2} \frac{R(\omega) - R(0)}{i\omega} e^{i\omega v} d\omega = 2 \sum_k a_k \operatorname{Si}[\omega_2(\tau_k - v)] - 2R(0)\operatorname{Si}(\omega_2 v), \tag{14}$$

where

$$\operatorname{Si} x = \int_0^x \frac{\sin x}{x} dx.$$

For sufficiently large values of ω_2, the function $\operatorname{Si}(\omega_2 x)$ is approximately a step function located at the point $x = 0$. The sharpness of the step function $\operatorname{Si}(\omega_2 x)$ increases with increasing ω_2. Each of the terms in Eq. (14) is a step, shifted by a distance τ_k along the ordinate.

The series $\sum_k a_k \operatorname{Si}[\omega_2(\tau_k - v)]$ will represent a curve which has steps at the points $v_k = \tau_k$ and which will be flat in the regions between these points.

$$\varphi_2(v) = \int_{-\omega_2}^{\omega_2} \frac{\operatorname{Im} R(\omega)}{i\omega} e^{i\omega v} d\omega = 2 \sum_k a_k [\operatorname{Si}(\omega_2(\tau_k - v)) - \operatorname{Si}(\omega_2(\tau_k + v))]. \tag{15}$$

$$\varphi_3(v) = \int_{-\omega_2}^{\omega_2} \frac{\operatorname{Re} R(\omega) - R(0)}{i\omega} e^{i\omega v} d\omega = 2 \sum_k a_k [\operatorname{Si}(\omega_2(\tau_k - v)) + \operatorname{Si}(\omega_2(\tau_k + v))] - R(0)\operatorname{Si}(\omega_2 v). \tag{15'}$$

The transformations in Eqs. (14) and (15) differ from those in (3), (7), and (8) in that they require an accurate knowledge of the frequency spectrum for $\omega \sim 0$, and do not require detailed information about R(ω) at high ω. Consequently, the two transforms have different ranges of applicability.

A group of transforms, using Eq. (15'), are shown in Fig. 5 for a case in which three waves interfere, with a varying time delay between the first two and a fixed time delay between the second two. The arrival time for each of the waves is determined by the position of the midpoint of the rise to the left of each extremal. These frequency spectrums were determined for a band from 10 to 70 cps. As is apparent, with this frequency band, correlation is possible between the wave components, even when the time shift between two neighboring waves is ≥0.012 sec.

In order to evaluate the effect of a narrow passband in seismic recording equipment on the form of the function given by Eq. (15), curves were computed for cases in which the lower

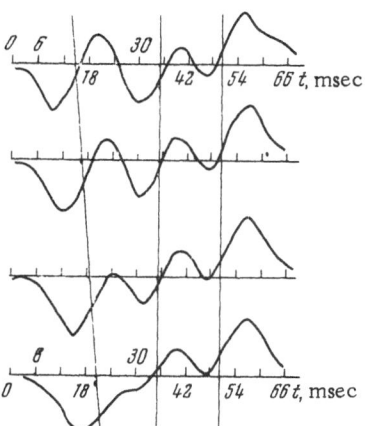

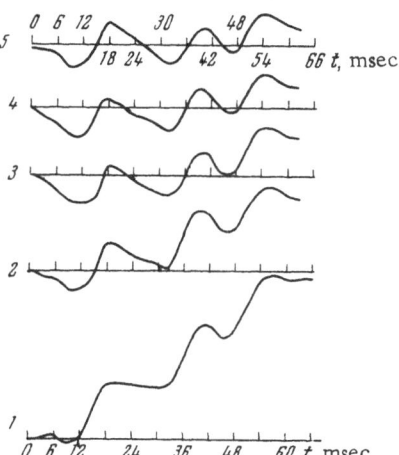

Fig. 5. Curves for $\varphi(v)$ for
interference between three
waves, with a pass band of
10-70 cps. The vertical
lines indicate the onsets
of the interfering waves.

Fig. 6. Curves for $\varphi(v)$ for
a variety of lower cutoff frequencies. 1) 0-100 cps; 2)
2-100 cps; 3) 4-100 cps; 4)
7-100 cps; 5) 10-100 cps.

cutoff frequency was set at 0, 2, 4, 7, or 10 cps, while the upper cutoff frequency was held at
100 cps (Fig. 6). It is apparent that if all frequencies down to zero are recorded, the curve
for $\varphi(v)$ will be a staircase function with the height of the steps being proportional to the amplitudes of the waves. As the lower frequencies are deleted progressively, each of the steps
becomes a maximum. Also, extremals which do not correspond to actual waves appear. An
essential feature, though, of the behavior of these curves is that the sharpness of the step corresponding to each of the waves remains practically the same. The arrival time for each wave
may be determined from the midpoint on the left side of the appropriate extremal in all cases.

Representation of Integral Transforms of the Frequency Spectrum
as Linear Filters Dependent on the Characteristics
of the Input Signal

Assuming that the variable of integration v in Eqs. (6), (7), (8), (11), (12), (14), and (15)
is the time t in (16), we find that the transformation processes in Eqs. (6), (7), (8), (14), and
(15) can be thought of as filters with characteristics depending on the properties of the incoming signal.

Taking v = t in Eq. (6)

$$B(\omega) = \begin{cases} S(\omega^{-1}) & |\omega| < \omega_2, \\ 0 & |\omega| > \omega_2, \end{cases}$$

and we have

$$S(t) = \int_{-\infty}^{\infty} B(\omega) A(\omega) e^{i\omega t} d\omega. \tag{16}$$

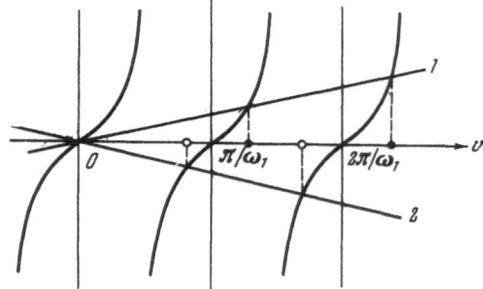

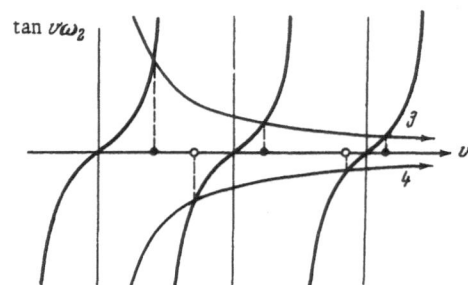

Fig. 7. Graphical solution of Eq. (22).
1, 2) Graphs of the function $y(v) = v/\tau_k \cdot$
$\cdot \tan \omega_2 \tau_k$; 3, 4) graphs of the function
$y = \tau_k/v \tan \omega_2 \tau_k$.

In order to represent the transformation processes in Eqs. (7) and (8) as linear filters, we use the relationships

$$2 \operatorname{Re} R(\omega) = R(\omega) + R(-\omega),$$
$$2i \operatorname{Im} R(\omega) = R(\omega) - R(-\omega).$$

Considering these expressions, we have

$$A_{\mathrm{Re}}(t) = \int_{-\infty}^{\infty} A(\omega) B(\omega) (1 - e^{2i(s_0(\omega)-a(\omega))}) e^{i\omega t} d\omega,$$

$$A_{\mathrm{Im}}(t) = -i \int_{-\infty}^{\infty} A(\omega) B(\omega) (1 + e^{2i(s_0(\omega)-a(\omega))}) e^{i\omega t} d\omega,$$

$$(17)$$

where $s_0(\omega)$ and $a(\omega)$ are the phase spectrums for the signals $\varphi(t)$ and $f(t)$.

Here, $A(\omega)$ is the spectrum of the input signal $\varphi(t)$, which may be converted with transforms of various types to functions which are amenable to geophysical interpretation. The basic property of all these transforms is that their characteristics — amplitude and phase — depend on the relationship between arrival times for the simple waves $a_k f(t - \tau_k)$ which comprise the complex wave form $\varphi(t)$.

Linear transforms with frequency spectrums having this property are not used for the separation of signals in time. However, the properties of an interference system are somewhat similar in concept with respect to the dependence of the frequency spectrum on the properties of the incoming signal. The frequency characteristic of an interference system depends on the direction of arrival of the waves.

Resolution of the Methods

The results given above for specific examples demonstrate the high degree of resolution which may be obtainable with these methods. However, these examples have not defined the limits of resolution, or the relation between resolution and other properties of the various transforms.

We will now obtain a numerical expression for the resolution of the transform $S(v)$. The resolution of $A_{\mathrm{Re}}(v)$ and $A_{\mathrm{Im}}(v)$ are quantitatively equivalent to that for $S(v)$. With $\omega_1 = \omega_2$, we obtain the following expression for $S(v)$:

$$S(v) = \sum_k a_k \frac{\sin 2\omega_1 (\tau_k - v)}{\tau_k - v}. \tag{18}$$

The individual waves correspond to single terms in Eq. (18):

$$S_k = a_k \frac{\sin 2\omega_1 (\tau_k - v)}{\tau_k - v}. \tag{19}$$

The zeros of the functions S_k are located at the points v_m, and the distance between them is $v_m - v_{m+1} = \pi/w_1$. The width of the principal maximum lobe of the function S_k is twice the distance between zeros, or $2\pi/\omega_1$. We will consider that the interfering waves are interpretable if the k-th principal maximum for the function $S(v)$ does not coincide with one of the points where $(k+1) = m$. In order for this not to happen, it is sufficient that the time separation between individual waves be no less than the width of each of the principal maximum lobes:

$$\tau_{k+1} - \tau_k \geqslant \frac{2\pi}{\omega_1}. \tag{20}$$

In comparing resolution for the transforms $A_{Re}(v)$, $A_{Im}(v)$, and $S(v)$, we can compare the width of the principal maximum lobes corresponding to the individual waves. The resolution will be better with narrower lobes. In all three transforms, the width Δv of a principal lobe corresponding to a single wave is twice the distance between zeros for the appropriate functions:

$$S_k = a_k \frac{\sin 2\omega_1 (\tau_k - v)}{\tau_k - v},$$

$$A_{Re,\,k} = a_k \frac{v \cos (\omega_2 \tau_k) \sin (v - \omega_2) - \tau_k \sin (\omega_2 \tau_k) \cos (v\omega_2)}{\tau_k^2 - v^2},$$

$$A_{Im,\,k} = a_k \frac{v \sin (\omega_2 \tau_k) \cos (v\omega_2) - \tau_k \cos (\omega_2 \tau_k) \sin (v\omega_2)}{\tau_k^2 - v^2}. \tag{21}$$

The zeros for the functions $A_{Re,\,k}$ and $A_{Im,\,k}$ are found as solutions to the following equations:

$$\tan v\omega_2 = \frac{\tau_k}{v} \tan \omega_2 \tau_k,$$

$$\tan v\omega_2 = \frac{v}{\tau_k} \tan \omega_2 \tau_k. \tag{22}$$

It is apparent from the graphical solution of the Eqs. (22) shown in Fig. 7 that for

$$\tan \omega_2 \tau_k > 0$$

the zeros for the function $A_{re,\,k}$ are closer together and the zeros for the function $A_{Im,\,k}$ are further apart than the comparable zeros for the function S_k.

On the other hand, if $\tan \omega_2 \tau_2 < 0$, the zeros for $A_{Im,\,k}$ are located closer together.

Thus, with $\tan \omega_2 \tau_k > 0$, the A_{Re} transform provides the narrowest principal lobe corresponding to the k-th wave, while with $\tan \omega_2 \tau_k < 0$, the A_{Im} transform provides the narrowest principal lobe. The width of the principal lobe is the same for all three transforms when $\tan \omega_2 \tau_k = 0$.

In evaluating the resolution of the transform $\varphi(v)$, we might consider that two waves are interpretable if the separation between the midpoints of the steps corresponding to each wave is at least three times the width of each of the steps (see Fig. 6). This condition is satisfied when

$$\omega_2 (\tau_{k+1} - \tau_k) \geqslant 3\pi. \tag{23}$$

It is apparent from Eqs. (20) and (23) that the resolution with the transform $\varphi(v)$ is somewhat less than the resolution with the transforms S, A_{Re}, and A_{Im}.

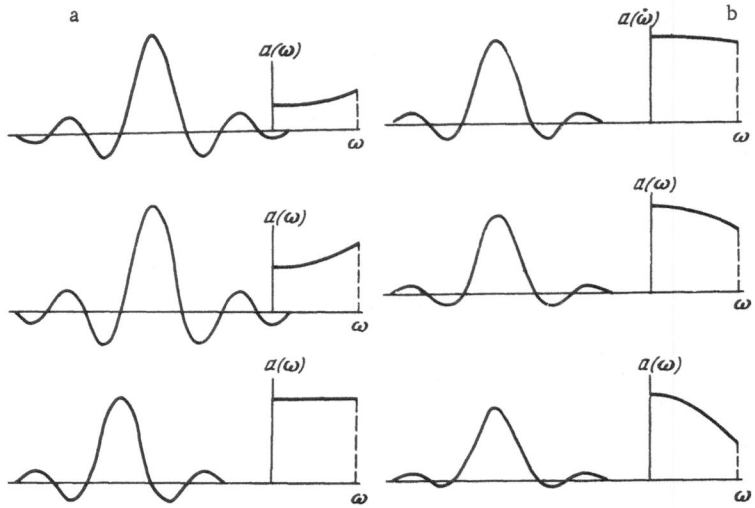

Fig. 8. The transform A_{Re} for a single wave. a) Curves for $A_{Re}(v)$ for a single wavelet, with the use of a distorted spectrum; b) ratio of the distorted spectrum to the true spectrum..

Stability of the Method with Respect to Variations

of the Interference Spectrums

These methods for separating wave fronts are based on the use of experimentally derived frequency spectrums:

$$R(\omega) = \frac{A(\omega)}{S(\omega)} = \sum_k a_k e^{-i\omega} \tau_k,$$

where $A(\omega)$ is the spectrum of the interfering wave packet, and $S(\omega)$ is the spectrum of each component of the interfering wave packet, taken to have a common form.

All of the above results were obtained on the basis of the supposition that $S(\omega)$ is known accurately. However, the spectrum $S(\omega)$ may differ because of errors or other factors, and these differences may not always be recognized. Let us examine how the properties of the transforms will change as a function of variations in $S(\omega)$.

Let the interfering waves have spectrums $S_k(\omega)$ and let the spectrum $S_0(\omega)$ be used in computing $R(\omega)$. Then the $R_H(\omega)$ which is obtained will be

$$R_H(\omega) = \frac{A(\omega)}{S_0(\omega)} = \sum_k \frac{S_k(\omega)}{S_0(\omega)} \cdot e^{-i\omega\tau}k. \tag{24}$$

We will assume that $a_k(\omega) = S_k(\omega)/S_0(\omega)$ is real. Then, each of the functions $a_k(\omega)$ may be expanded in a Fourier cosine series over the interval for which $R_H(\omega)$ is given:

$$a_k(\omega) = \sum_l a_{k,l} \cos \frac{\pi}{\omega_2} l\omega, \tag{25}$$

where ω_2 is the upper frequency cutoff during recording.

We obtain the following for $R_H(\omega)$ for Eqs. (24) and (25):

$$R_H(\omega) = \frac{1}{2} \sum_k \sum_l a_{k,l} \left[e^{-i\omega \left(\tau_k - \frac{\pi}{\omega_2} l \right)} + e^{-i\omega \left(\tau_k + \frac{\pi}{\omega_2} l \right)} \right]. \qquad (26)$$

If we apply the transformations given above to $R_H(\omega)$, then each term in the Fourier series (25) will generate a pair of principal maximums or two steps, shifted with respect to each other. Thus, if the coefficients $a_{k,l}$ drop off slowly with l, the general picture becomes highly complicated.

However, for cases in which $a_k(\omega)$ vary slowly and the coefficients $a_{k,l}$ drop off quickly with increasing l, the most important terms are those containing, $a_{k,0}$. The terms in Eq. (26) containing the coefficient $a_{k,0}$ have the same phase shift as the phase shift between the interfering waves. As a consequence, the characteristic points on the curves for each of the transformations corresponding to the wave arrivals may be picked more closely.

Figure 8 shows a group of transforms $A_{Re}(v)$ for a single wavelet, where the curve $a_1(\omega)$ is a monotonic function for $\omega > 0$. It is apparent that the form of the curves for $A_{Re}(v)$ is changed only slightly.

Results

1. The problem of decomposing a group of waves $\varphi(t)$ consisting of several simple waves $a_k f(t - \tau_k)$ and determining the positions of the individual wave fronts when they interfere with each other may be reduced to the problem of determining the amplitudes of the simple waves and their time delays relative to some arbitrary zero time based on the knowledge of the complex spectrum of the wave group.

2. The complex frequency spectrum $R(\omega)$ for a group of waves is determined as being made up of spectral components in the same manner as the complex wave form $\varphi(t)$ is made up of simpler waves.

3. In determining the amplitudes a_k and phase shifts τ_k of the simple waves, it is necessary to apply a Fourier transformation to $R(\omega)$ or some function of $R(\omega)$. The values for a_k and τ_k are determined from diagnostic points on the curves for the Fourier transforms.

4. The functions which may be selected for the Fourier transformation include 1) Im $R(\omega)$; 2) Re $R(\omega)$; 3) Im $R(\omega)/i\omega$; 4) [Re $R(\omega) - R(0)]/i\omega$.

Because diagnostic points which are not actually associated with real wave components may appear on these plotted curves for the Fourier transforms and because these false points occur differently on the various curves, a comparison of the Fourier transforms for several of these functions improves the reliability of the results.

5. The resolution of the methods is higher with a broader frequency band over which $R(\omega)$ has been determined. The quality of results is improved markedly if the lower frequency cutoff is made approximately zero.

LITERATURE CITED

1. I. S. Berzon, A. M. Epinat'eva, G. N. Pariiskaya, and S. P. Starodubrovskaya, Dynamic Characteristics of Seismic Waves in Real Media, Izd. Akad. Nauk SSSR, 1962.
2. L. L. Khudzinskii, On determining some properties of layers of moderate thickness from the spectrum of reflected waves, Izv. Akad. Nauk SSSR, Ser. Geofiz., No. 5 (1961).
3. M. E. Grin', Determination of the time delay of waves in an interference zone by use of spectral analysis, Dokl. Akad. Nauk UkSSR, No. 10 (1959).

INTERPRETATION OF MULTIDIMENSIONAL SPECTRUMS FOR SEISMIC WAVES

S.A. Kats

The one-dimensional Fourier transform

$$f(t, x, y, z) = \frac{1}{2\pi} \int_{-\infty}^{\infty} S(\omega, x, y, z) e^{i\omega t} d\omega, \tag{1}$$

$$S(\omega, x, y, z) = \int_{-\infty}^{\infty} f(t, x, y, z) e^{-i\omega t} dt. \tag{1'}$$

is used in the analysis of a number of seismic problems where the frequency spectrum is of interest [1, 3]. In addition to the one-dimensional Fourier transform, it is sometimes advantageous to make use of a multidimensional Fourier transform, particularly for describing the properties of seismic waves which are functions of several variables. The following function may be defined formally as the k-th degree Fourier transform of a function of n variables:

$$f(x_1, \ldots, x_n) = \frac{1}{(2\pi)^k} \int_{-\infty}^{+\infty} \cdots \int_{-\infty}^{+\infty} B(\omega_{x_1}, \ldots, \omega_{x_l}, x_{k+1}, \ldots, x_n) e^{i \sum_{l=1}^{l} \omega_{x_l} x_l} d\omega_{x_1}, \ldots, d\omega_{x_l},$$

$$B(\omega_{x_1}, \ldots, \omega_{x_k}, x_{k+1}, \ldots, x_n) = \int_{-\infty}^{+\infty} \cdots \int_{-\infty}^{+\infty} f(x_1, \ldots, x_n) e^{-i \sum_{l=1}^{l} \omega_{x_l} x_l} dx_1, \ldots, dx_l. \tag{2}$$

Here, x_l is the l-th variable and ω_{xl} is the l-th frequency. The notation $S(\omega, x, y, z)$ will be used here to represent the one-dimensional spectrum of the seismic waves, and the notation $B(\omega_{x_1}, \ldots, \omega_{x_k} x_{k+1}, \ldots, x_n)$ will be used to represent the Fourier spectrum of k-th degree. It is obvious that

$$B(\omega_{x_1}, x_2, \ldots, x_n) = S(\omega, x_2, \ldots, x_n).$$

In analogy with the one-dimensional running spectrum

$$S(\omega, \hat{T}, x, y, z) = \int_{0}^{T} f(t, x, y, z) e^{i\omega t} \partial t$$

we may define the multidimensional running spectrum in the form

$$R(\omega_{x_1}, \ldots, \omega_{x_l}, T_1, \ldots, T_k, x_{k+1}, \ldots, x_n) = \int_0^{T_1} \cdots \int_0^{T_k} f(x_1, \ldots, x_n) e^{-i \sum_{l=1}^{k} x_l \omega_{x_l}} dx_1, \ldots, dx_l. \tag{3}$$

In this paper, we will examine the relationship between the multidimensional and one-dimensional running spectrums for seismic waves. As will be shown below, these relationships may be used in interpreting seismic wave behavior.

Multidimensional Spectrum of Plane Waves

We may represent a plane dispersive wave as

$$f(t, x, y, z) = \frac{1}{2\pi} \int_{-\infty}^{\infty} S(\omega) e^{i\omega \left(t + \frac{\alpha x + \beta y + \gamma z}{v(\omega)} \right)} d\omega, \tag{4}$$

where $v(\omega)$ is a complex function of frequency if the wave is propagating in an absorptive medium. The inverse Fourier transform of Eq. (4) gives the following result:

$$S(\omega) e^{+i\omega \frac{\alpha x + \beta y + \gamma z}{v(\omega)}} = \int_{-\infty}^{\infty} f(t, x, y, z) e^{-i\omega t} dt. \tag{5}$$

The two-dimensional spectrum for the wave given by Eq. (4) with x taken as a transform variable will have the following form:

$$B(\omega, \omega_x, T_2, y, z) = \int_{-\infty}^{+\infty} \int_0^{T_2} f(t, x, y, z) e^{-i(\omega t + \omega_x x)} dt\, dx. \tag{6}$$

Equation (6) is integrated first with respect to t and then with respect to x, so that with consideration of Eq. (5) we obtain

$$B(\omega, \omega_x, T_2, y, z) = S(\omega) e^{-i\omega \frac{\gamma z + \beta y}{v(\omega)}} \cdot e^{-i\left(\frac{\omega \cdot \alpha}{v(\omega)} - \omega_x \right) \frac{T_2}{2}} \frac{\sin\left[\left(\frac{\omega \cdot \alpha}{v} - \omega_x \right) \frac{T_2}{2} \right]}{\frac{\omega \cdot \alpha}{v} - \omega_x}. \tag{7}$$

Similarly, for the three-dimensional running spectrum we have

$$B(\omega, \omega_x, \omega_y, T_2, T_3, z) = S(\omega) \exp\left(-i\omega \frac{\gamma z}{v(\omega)} \right) e^{-i\left(\frac{\omega \cdot \alpha}{v} - \omega_x \right) \frac{T_2}{2} - i\left(\frac{\omega \beta}{v} - \omega_y \right) \frac{T_3}{2}} \times$$

$$\times \frac{\sin\left[\left(\frac{\omega \cdot \alpha}{v} - \omega_x \right) \frac{T_2}{2} \right]}{\frac{\omega \cdot \alpha}{v} - \omega_x} \cdot \frac{\sin\left[\left(\frac{\omega \beta}{v} - \omega_y \right) \frac{T_3}{2} \right]}{\frac{\omega \beta}{v} - \omega_y}. \tag{8}$$

We have from Eqs. (7) and (8)

$$B(\omega, \omega_x, \omega_y, T_2, T_3, z) = B(\omega, \omega_x, T_2, y, z) e^{i\omega \frac{\beta y}{v}} \cdot e^{-i\left(\frac{\omega\beta}{v} - \omega_y\right)\frac{T_3}{2}} \cdot \frac{\sin\left[\left(\frac{\omega\beta}{v} - \omega_y\right)\frac{T_3}{2}\right]}{\frac{\omega\beta}{v} - \omega_y}. \tag{9}$$

Here ω_x and ω_y are spatial frequencies with the dimensions m^{-1}. We will examine the advantages of the two-dimensional Fourier transform. With the continuous profiling method of acquiring seismic data, it is possible to obtain both the one-dimensional and two-dimensional spectrums.

Inverting Eq. (6), we obtain [2]:

$$f(t, x, y, z) = \frac{1}{(2\pi)^2} \int_{-\infty}^{\infty} \int_{-\infty}^{\infty} B(\omega_1, \omega_x, T_2, y, z) e^{i(\omega t + \omega_x \cdot x)} d\omega \cdot d\omega_x,$$

or, writing $v = \omega/\omega_x$,

$$f(t, x, y, z) = \frac{1}{(2\pi)^2} \int_{-\infty}^{\infty} \int_{-\infty}^{\infty} \frac{1}{\omega} B\left(\omega, \frac{\omega}{v}, T_2, y, z\right) e^{i\omega\left(t + \frac{x}{v}\right)} d\omega dv. \tag{10}$$

Here, v is a parameter having the significance of the phase velocity for various harmonics. If $f(t, x) = f(t - x/v)$, then

$$f(t, x) = \frac{1}{2\pi} \int_{-\infty}^{\infty} S(\omega) e^{i\omega\left(t - \frac{x}{v}\right)} d\omega. \tag{11}$$

Comparing the one-dimensional spectrum (11) and the two-dimensional spectrum (10) for a wave propagating without change in form, we note the following: the one-dimensional Fourier integral decomposes the wave into harmonic components with differing frequencies but a common phase velocity. The two-dimensional Fourier integral decomposes the wave into harmonic components with differing frequencies and differing phase velocities. For dispersive waves, the one-dimensional Fourier integral gives a decomposition of the waves in which each frequency corresponds to a single harmonic with a specific phase velocity which is related to this frequency. The two-dimensional Fourier integral gives a decomposition of the waves in which each frequency corresponds to an infinite variety of harmonics with phase velocities in the range $(-\infty, +\infty)$.

Let a seismic wave $f(t, x)$ be recorded along a profile x, with $y = z = 0$. Then, from (7) we find that in the (ω, ω_x) plane there is a line, given by the equation

$$\frac{\omega \cdot \alpha}{v(\omega)} - \omega_x = 0, \tag{11'}$$

along which

$$B(\omega, \omega_x, T_2, y, z) = S(\omega) \frac{T_2}{2}. \tag{12}$$

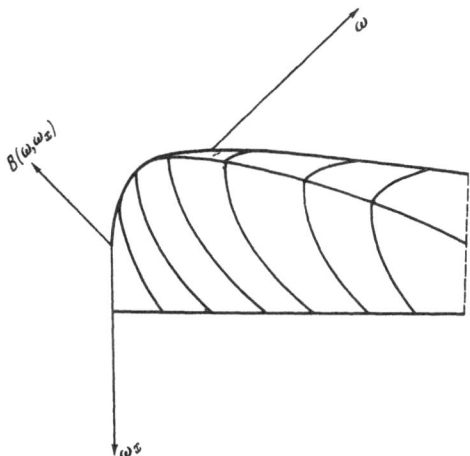

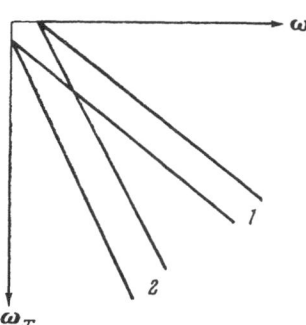

Fig. 1. Principal maximum of a two-dimensional wave spectrum when the phase velocity is independent of frequency.

Fig. 2. Regions of principal maximums (1, 2) for two-dimensional spectrums of waves with different phase velocities.

The value for ω_x which satisfies Eq. (11') will be designated as $\bar{\omega}_x$. Considering that

$$e^{-i\left(\frac{\omega \cdot \alpha}{v} - \omega_x\right)\frac{T_2}{2}}\ \frac{\sin\left(\frac{\omega \cdot \alpha}{v} - \omega_x\right)\frac{T_2}{2}}{\frac{\omega \cdot \alpha}{v} - \omega_x} = \int_0^{T_2} e^{-i\left(\frac{\omega \cdot \alpha}{v} - \omega_x\right)x}\,dx,$$

we find that away from the surface specified by Eq. (11'),

$$|B(\omega, \omega_x, T_2, y, z)| < |S(\omega)|\frac{T_2}{2} = |B_2(\omega, \bar{\omega}_x, T_2, y, z)|. \tag{13}$$

The surfaces specified by Eq. (11) will be situated in four-dimensional complex space, if $v(\omega)$ is complex, where the coordinates are Re ω, Im ω, Re ω_x, and Im ω_x.

It is apparent from Eqs. (12) and (13) that in the region of maximum values for the modulus of the two-dimensional spectrum, the two-dimensional spectrum will coincide with the one-dimensional spectrum, except for a multiplying constant. The form of the principal maximum in the amplitude for the two-dimensional spectrum $|B(\omega, \omega_x, T_2, y, z)|$ is shown in Fig. 1 for a wave in which the phase velocity is independent of frequency. The region of maximum values of $|B(\omega, \omega_x, T_2, y, z)|$ in the (ω, ω_x) plane is determined by the phase velocity of the wave, and will differ for waves with different phase velocities. The regions of the principal maximums in the two-dimensional amplitude spectrums of two-waves with different phase velocities are indicated in Fig. 2. These regions overlay at $\omega \approx 0$, and separate at higher frequencies.

Similarly, it may be shown that in the region of maximum values of amplitude in a three-dimensional spectrum, the three-dimensional spectrum coincides with the one-dimensional spectrum, except for a multiplying constant.

We will now examine the possibility of using two-dimensional spectrums for interpreting waves in an interference region.

Let an interference pattern $\varphi(t, x)$ be recorded along the profile $x_1 < x < x_2$:

$$\varphi(t, x) = \sum_m f_m(t, x),$$

where $f_m(t, x)$ is a wave, given in the form

$$f_m(t, x) = \frac{1}{2\pi} \int_{-\infty}^{\infty} S_m(\omega)\, e^{i\omega\left(t - \frac{x}{v_m}\right)}\, d\omega.$$

The two-dimensional spectrum for the wave pattern $\varphi(t, x)$ will be

$$B(\omega, \omega_x, y, z) = \sum_m S_m(\omega)\, e^{-i\left(\frac{\omega}{V_m} - \omega_x\right)\frac{x_1 + x_2}{2}} \cdot \frac{\left[\left(\frac{\omega}{v_m} - \omega_x\right)\frac{x_1 - x_2}{2}\right]}{\left(\frac{\omega}{v_m} - \omega_x\right)}.$$

Each of the waves has its own region of maximum values in the (ω, ω_x) plane if the phase velocities of the wave components differ by a sufficient amount.

If a region of maximum values may be recognized in the two-dimensional spectrum corresponding to the m-th wave component, its phase velocity at a frequency ω is given by

$$v_m(\omega) = \frac{\omega}{\bar{\omega}_x}.$$

Here, ω and $\bar{\omega}$ are the coordinates of the point in the frequency domain where $|B(\omega, \omega_x, y, z)|$ passes through a maximum for a fixed value of ω. Curved waves will be dispersed if $v_m(\omega)$ is the case for a set of frequency ω.

In evaluating the resolution obtainable in this procedure, we may consider that two waves are resolvable if the regions of principal maximum for them overlay by no more than 50%. This means that in order for the individual terms in Eq. (14) not to coalesce, it is necessary that

$$\omega \frac{x_1 - x_2}{2}\left(\frac{1}{v_m} - \frac{1}{v_l}\right) > \pi$$

or

$$\sigma\Delta t > \frac{1}{f},$$

where $\sigma\Delta t$ is the difference in timeshift of the harmonics with frequency ω along the transform base, and f is frequency in cps.

It is apparent from this that resolution increases with increasing frequency.

Results

1. The problem of interpreting waves which differ in phase velocity in an interference pattern may be reduced to a problem of interpreting multidimensional spectrum for the interfering waves.

2. It is reasonable to make use of multidimensional spectrums when waves with changing form must be interpreted from an interference pattern.

3. The resolution possible with the method increases with larger geophone spacings and higher frequencies used in determining phase velocities.

LITERATURE CITED

1. I. S. Berzon, A. M. Epinat'eva, G. N. Pariiskaya, and S. P. Starodubrovskaya, Dynamic characteristics of seismic waves in real media, Izd. Akad. Nauk SSSR, 1962.
2. V. L. Dutkin and A. P. Prudnikov, Operational Computations in Two Variables and Their Application, Fizmatgiz, 1964.
3. L. L. Khudzinskii, Frequency analysis of interference patterns of seismic waves, Tr. Inst. Fiz. Zemli, No. 6:173 (1959).

PHYSICAL BASIS FOR THE USE OF REFLECTED WAVE FORMS FOR STUDYING GROUPS OF THIN BEDS

S.P. Starodubrovskaya, T.S. Makushkina, and E. V. Vilkova

Introduction

At present, considerable thought is being given to the study of reflections caused by groups of thin beds. Electric and acoustic well logs described earlier in this collection have shown that the occurrence of thin-bedded sequences is common. The bed thickness may be comparable to the lengths of the waves travelling through them in some cases, or considerably longer or shorter in other cases. In view of this, there is both scientific and practical interest in the study of wave propagation in groups of layers with varying thickness. The study of groups of layers with varying thickness is of particular practical importance in the exploration for and development of oil and gas reserves, which occur preferentially in the permeable beds. The greater part of oil and gas reserves is associated with sequences of detrital rocks, occurring at various depths below the surface (for example, the fields on the Russian platform, in the Krasnodar Belt, and others). The detrital rocks in which oil and gas occur are alternating beds of sandy and shaly facies (siltstones and shales). From the seismologists' point of view, such rocks may be represented as sequences of periodically alternating beds with nearly the same velocity profile in them. In the present paper, we will examine a seismic representation which approximates a real sequence of detrital rocks and study the dynamic wave properties (the wave form) for waves reflected from such sequences which vary in thickness. This is done using data from theoretical studies.

§1. Seismic Representation of the Actual Medium

Little is known at present about the structure of sequences of detrital rocks. Some information about the structure of various sequences of detrital rocks has been obtained by the Institute of Physics of the Earth of the Academy of Sciences of the USSR and by other groups, on the basis of acoustic logging data from the Russian platform and the Krasnodar Belt, as described in the first part of this book, as well as on the basis of electric logging data, standard seismic logs, and the results of studies on speeds on propagation of elastic waves in samples. Summarizing the available data on the seismic structure of sequences of detrital rocks, we may draw the following conclusions about the properties of such sequences.

1. Detrital sequences are alternating beds with high speeds of propagation ($V_2 = 2900$–4000 m/sec in sandstones and siltstones) and low speeds ($V_1 = 2600$–3500 m/sec in shales and mudstones). Thus, the speeds in the rocks comprising a detrital sequence are close to one another. The ratio of speeds at the boundaries of beds in a sequence is $V_1/V_2 = 0.7$–0.9 on the average; that is, there is relatively little differentiation in speed of propagation between the layers.*

---

* Cases are known (on the Russian platform) of the presence of carbonate interbeds with $v = 4500$ m/sec in detrital sequences.

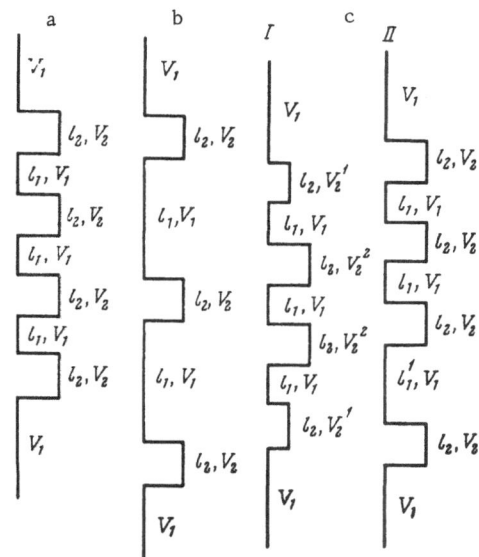

Fig. 1. Representations of seismic properties in sequences of detrital rocks taken for this study. a) Periodic sequence with equal values for τ; b) sequence which is periodic in thickness and speed of propagation; c) nonperiodic sequences: I) periodic in thickness but not in speed of propagation; II) periodically-layered portion of beds in a sequence.

2. The following variants of alternating sequences of layers are most likely to occur:

a) A periodic alternation of beds with high and low velocities; that is, the thicknesses (l_1 and l_2) and the speeds of propagation (V_1 and V_2) of the beds are fixed. The case where $l_1 \gg l_2$ is encountered more commonly than the reverse; that is, the thickness of the shale interbeds is usually much greater than the thickness of the sandstone interbeds. The theoretical computations, we have considered to be model periodic sequences in which the ratio of transit times for longitudinal waves, $\tau_1/\tau_2 V_1$ is greater than 1 where the transit times times V_1 and $\tau_2 = 2l_2/V_2$ represent the two-way travel tines in the beds with low and high velocity, respectively (see Fig. 1). The ratio τ_1/τ_2 will be designated as S.

For $S = 1$ (that is, $\tau_1 = \tau_2 = \text{const}$) the transit time is the same in each bed and we have the case of "periodicity with equal τ" (Fig. 1a). In cases where there is little contrast in speed of propagation between beds (as for example, with $V_1/V_2 \approx 0.9$, as it is in the Krasnodar Belt), the beds in a sequence will also have about the same thicknesses, $l_2/l_1 \approx 1$.

For $S > 1$, the transit time for waves in the low-speed beds is larger than the corresponding transit time in the high-speed beds ($\tau_1 > \tau_2$); that is, we have periodicity in thickness and propagation speed with different fundamentals (Fig. 1b). The most likely cases in actual practice are those in which the value for S ranges from 1 to 10-15; with $V_1/V_2 \approx 0.9$, this range of values for S corresponds to a range of values of 1-0.1 for the ratio of bed thicknesses for the high-speed (l_2) and low-speed (l_1) members; that is, the bed thicknesses may differ by an order of magnitude.

b) A nonperiodic alternation of layers with high and low propagation speeds: 1) a high-speed layer at the edge of a sequence or within a sequence which has a speed different than that of the rest of the high-speed layers in the group; and 2) the thicknesses of the high-speed layers in a sequence vary. In the theoretical calculations we have considered nonperiodic sequences of layers in which $\tau_1 \neq \tau_2$; that is, both propagation speeds and thicknesses are nonperiodic (Fig. 1c). The number of possible variants in a nonperiodic sequence is very large. In the Krasnodar Belt, for example, two types of nonperiodicity are common: periodicity is exhibited by bed thickness but not by propagation speeds or, in other cases, periodicity is noted over a part of a sequence (Fig. 1c).

3. The medium adjoining a sequence of productive elastic layers varies from place to place. In some cases, it consists of dense carbonates (as for example, on the Russian Platform, with $V \approx 5000$ m/sec), and in other cases, it consists of relatively homogeneous shale beds (as for example, the shales in the Krasnodar Belt, with $V = 2600$ m/sec). Thus, in some cases (the Krasnodar Belt), the contrast in propagation speeds between a thin-bedded sequence and the surrounding medium is about the same as the contrasts within the sequence, and the entire section has only weak contrasts in speeds of propagation. In other cases (the Russian

platform), the contrast in speed of propagation between a thin-bedded sequence and the surrounding medium is quite different than the contrasts within the sequence, and as a result, there are stronger contrasts in speeds of propagation for the section as a whole ($\widetilde{V}_1/V_2 = 1.3$) than there are within the thin-bedded sequence ($V_1/V_2 = 0.9$).

In the present study, we have computed reflected wave forms for all of these types of seismic structures. We have determined frequency spectrums for the sequences of thin beds which reflect seismic waves and the wave forms for waves reflected from thin-bedded sequences with a variety of structures and with various numbers of beds. We have studied periodic sequences of layers with values of S ranging from 1 to 14, in which the contrast in propagation speed with the surrounding medium was the same as those within the thin-bedded sequence, with $V_1/V_2 = 0.7$-0.9, with the density being the same in all beds, with the number of beds in a sequence ranging from 3-7 and the number of high-speed beds being $p = 2$-4; attenuation within the beds was not considered.

We have considered sequences with varying thickness; the ratio of the total thickness l_Σ to the average wavelength in them l_{av} was varied from 0.1 to 5-6. In addition, we considered several types of nonperiodic sequences (see §6). The parameter l_Σ/l_{av} may be expressed as the product of the transit time through all of the beds τ_Σ at the dominant frequency f_d in the incident pulse. In addition, the average wavelength may be specified by dividing the average propagation speed $\overline{V}$ in the sequence by the dominant frequency f_d ($\lambda_{av} = \overline{V}/f_d$). The transit time τ_Σ through the entire sequence may be computed from the total thickness of the sequence l_Σ and the average velocity $\overline{V}$ [1]:

$$\tau_\Sigma = \frac{l_\Sigma}{\overline{V}}.$$

In this paper, we present data on the wave forms for waves reflected from a periodic sequence of beds, and consider the possibility of determining, from the properties of the reflected waves, the structure (the number of layers and the nature of the periodicities) of a sequence which is varying in thickness.

§2. Method Used in the Theoretical Studies

In the theoretical studies, we computed synthetic seismograms, spectrums for reflected waves, and the spectral characteristics for sequences of layers for wave reflection.

The synthetic seismograms were computed using either the digital computer program of N. G. Mikhailova and B. S. Pariiskii [1], or, in the case of small contrasts in speed of propagation (with $V_1/V_2 \approx 0.9$), by summing singly reflected waves from the top and bottom of each bed, with consideration of travel times, but without consideration of refraction along boundaries. The computations show that in the case of weak contrasts in speed of propagation at boundaries between beds, neglecting refractions along boundaries and multiple reflection of waves has practically no effect on the recorded form of the reflected waves [1]. The synthetic seismograms were computed for a single form of experimental pulse, which was well approximated by a theoretical pulse of the form

$$F(t) = A_0 e^{-\beta^2 t^2} \sin \omega_0 t, \qquad \frac{\beta}{\omega_0} = \frac{1}{7} \quad \text{or} \quad \frac{1}{8}.$$

The frequency spectrums for reflected waves were determined on a frequency analyzer or on the BESM-2 computer.

The spectral characteristics were computed for the periodic and nonperiodic sequences of layers on the BESM-2 computer using a program [1] which gives an exact solution to the problem for reflection from a thin sequence of layers. A few of the spectral characteristics were obtained using an approximate expression given in reference [2], in which multiple reflections and refractions along boundaries are not specifically considered.

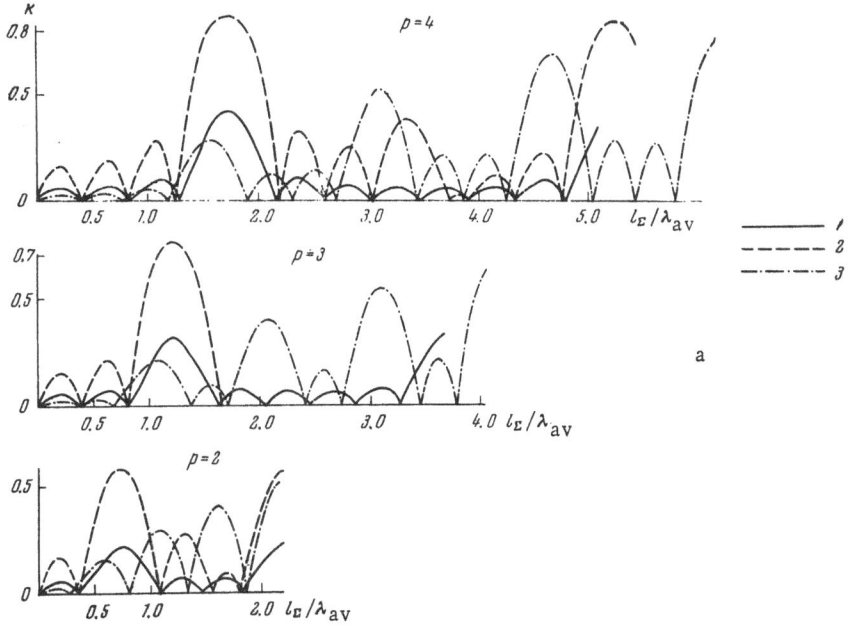

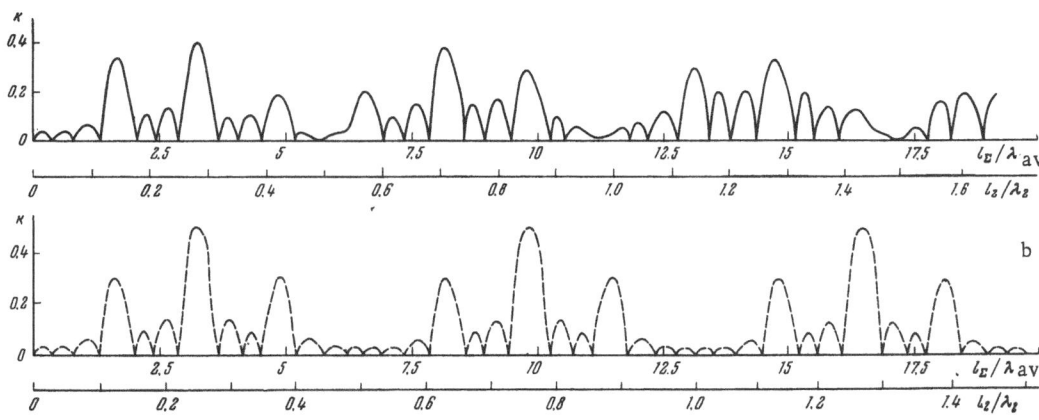

Fig. 2. Spectral characteristics for reflection of waves from a sequence of beds for various values of S. a) Spectral characteristic $K = K(l_\Sigma/\lambda_{av})$ for various values of S and p: 1) S = 1; 2) S = 1.4; 3) S = 14; the case S = 1 was computed for V_1/V_2 = 0.9, and the cases S = 1.4 and 14 were computed for V_1/V_2 = 0.7. b) Comparison of spectral characteristics $K = K(l_\Sigma/\lambda_{av})$ for a single value, p = 4, and with various values of S: 1) S = 2.5; 2) S = 3.

§3. Spectral Characteristics of Sequences with Various Periodicities in Thin Layers

As noted in reference [2], the spectral characteristics of a sequence may be represented as the product of two periodic terms, independently of the nature of the periodicity. The shape of the characteristic; that is, the positions of the extremals along the l_Σ/λ_{av} axis (see Fig. 2a), is determined by the value for S. The value for $l_{\Sigma_1}/\lambda_{\Sigma_1}$ is a parameter describing the repeating elements in a sequence made up of layers with high speed (l_2) and low speed (l_1); $l_{\Sigma_1} = l_1 + l_2$

and $\lambda_{\Sigma_1} = \overline{V}_1/t_d$; $\overline{V}_1$ is the average propagation speed in an element of a sequence; and f_d is the dominant frequency in the incident pulse. For $S = 1, 2, \ldots,$ or any integer number, the spectral characteristic of a sequence is a periodic function (Fig. 2b, the case $S = 3$); that is, all the properties of internal modulation in a periodicity of l_2/λ_2 are repeated periodically with a period $0.5\, l_2/\lambda_2$. With S not an integer, the spectral characteristic will be a quasiperiodic function — the nature of the integral modulation within the periodicity of l_2/λ_2 varies from period to period, as shown in Fig. 2b for the case $S = 2.5$. Thus, it is difficult to predict the internal modulation in the form of a spectral characteristic beforehand.

An analysis of the spectral characteristics of a periodic sequence has been given in references [1, 2]. The value for l_Σ/λ_{av} is related to l_2/λ_2 (to the properties of the layers with the least transit time) by the following relationship:

$$l_\Sigma/\lambda_{av} = [p + S(\mu - 1)]l_2/\lambda_2;$$

and the value for $l_{\Sigma_1}/\lambda_{\Sigma_1}$ is expressed in terms of l_2/λ_2 as follows:

$$l_{\Sigma_1}/\lambda_{\Sigma_1} = l_2/\lambda_2(1 + S); \qquad l_\Sigma/\lambda_{av} = l_{\Sigma_1}/\lambda_{\Sigma_1}(p-1) + l_2/\lambda_2.$$

Thus, knowing the transit time in a single layer $\tau_2 = 2l_2/V_2$ or l_2/λ_2, it is possible to determine the scaling factor l_Σ/λ_{av} for sequences with various numbers of layers and various periodicities.

For convenience, we will classify sequences according to thickness. In analogy to the case of a thin homogeneous layer, sequences of layers may be classified according to the ratio of the total thickness l_Σ to the average wavelength in the sequences λ_{av} as very thin, thin, intermediate, or thick:

1. Very thin sequences are characterized by $l_\Sigma/\lambda_{av} < 0.3$;

2. thin sequences have values $l_\Sigma/\lambda_{av} > 0.3$ but less than the value for which the first principal extremal appears;

3. sequences of intermediate thickness, characterized by

$$(l_\Sigma/\lambda_{av})_{pr} < l_\Sigma/\lambda_{av} < 1.5\,[p + S(p - 1)],$$

where $(l_\Sigma/\lambda_{av})_{pr}$ is the value for l_Σ/λ_{av} corresponding to the principal maximum of the characteristic; and

4. thick sequences are characterized by

$$l_\Sigma/\lambda_{av} > 1.5\,[p + S(p - 1)].$$

The class of very thin layers includes sequences, as in the case of homogeneous layers, for which the interference between reflected waves provides virtually no expression in the dynamic characteristics for these waves.

In the case of thin sequences, the reflections from the top and bottom of the sequence are not separated on recording. The recorded wave form either bears a complex interference form for $l_\Sigma/\lambda_{av} < (l_\Sigma/\lambda_{av})_{pr}$ or a simplified form for $l_\Sigma/\lambda_{av} \approx (l_\Sigma/\lambda_{av})_{pr}$.

With $l_\Sigma/\lambda_{av} > (l_\Sigma/\lambda_{av})_{pr}$, as in the case of sequences of intermediate thickness, satisfactory wave forms are not obtained with any value of l_Σ/λ_{av}. With large values for l_Σ/λ_{av}, the recorded wave form has either a quasiresonant form or easily resolved waves from the top and bottom of the sequence.

A thin sequence is considered to be a sequence for which the waves from the top and bottom of the sequence are separated by the least resolvable transit time for that particular case. These values for l_Σ/λ_{av} (> 3) are very large, corresponding to sequence thicknesses of several hundred meters. Below, we will consider the spectral response function K (l_Σ/λ_{av}) from the point of view of understanding the effects of the following factors: 1) variations in the periodicity of layers in the sequence; and 2) variations in the number of layers in a sequence while the periodicity is unchanged.

Effect of Variations in the Periodicity of Layers in a Sequence on its Reflection Spectrum

We will now consider the effect of changes in the periodicity of layers within a sequence for various specific values of total sequence thickness, with the total number of layers remaining constant. Calculations have been carried out for sequences containing 3, 5, or 7 layers. The value for S in the sequence varied from 1 to 14.

Consideration of the spectral response of sequences computed for the various numbers of beds (3-7) and various values for S (1-14) has shown that the relationship of the form of the spectral response function to the value of S is the same for sequences with different numbers of layers (see Fig. 2). Therefore, further attention was given to the nature of the variation of the form of the spectral response caused by changes in S, independent of the number of beds comprising the sequence.

With increasing S (an increase in S corresponds to a decrease in l_2/l_1 if V_1/V_2 is const), the amplitudes of the spectral components in the response function decrease and the periodicity of the maximums in the function decrease (see Fig. 2). The extremal points are characteristically shifted towards lower values for l_Σ/λ_{av}. The greatest reduction in period is observed for sequences with the least number of layers.

The dependence of the form of the spectral response on the character of periodicity of the layers within the sequence varies over different ranges of sequence thickness (Fig. 2).

In the case of very thick or very thin sequences, the form of the spectral response function is practically unchanged as S is varied, independently of the number of layers in the sequences; that is, the number of extremal points, the ratios of amplitudes of extremal points and even the positions of the extremal points on the l_Σ/λ_{av} axis are unchanged.

In the case of a sequence of moderate thickness, the form of the spectral response function depends on the periodicity of the layers in the sequence. With increasing S (S > 1), for sequences with the same l_Σ/λ_{av}, the number and amplitude of the principal and subsidiary maximums increase, the ratios of amplitudes for the maximums changes, and the maximums shift toward higher frequencies (Fig. 2). These changes are controlled by the shift of the first minimum in the modulating periodicity to a larger l_Σ/λ_{av} with increasing S [2].

These changes in the spectral response function are observed for any value of l_Σ/λ_{av} within the first modulation period if the value for S is varied from 1 to 1.5. With a further increase in S, there is a range of values for l_Σ/λ_{av} for which the spectral response function is insensitive to variations in S. For example, for a sequence having a thickness falling within the range of values for l_Σ/λ_{av} between the first and second maximum points, a change in S from 1.5-2.5 does not affect the spectral response function for the sequence.

Effect of a Change in the Number of Layers in a Sequence on Its Spectral Response

We also investigated the effect of the variation of the number of layers in a sequence from 3-7 with a fixed value for S, on the spectral response function, for several specific values of sequence thickness. We considered sequences with several fixed values for S. Consistent changes in the spectral response function were noted as the number of layers in the sequence was varied, independently of the nature of the periodicity of the layers in the sequence (independently of the value for S) [1]. With an increase in the number of layers in a sequence, but with a specified value for l_Σ/l_{av}, the number of subsidiary maximums between the principal maximums of the response function increased [1, 2] (Fig. 2); that is, the absolute values for the amplitudes of the response function were increased and the period between maximums was lengthened. Because of this latter behavior, a particular maximum in the response function for a sequence shifts towards higher frequencies as n is increased. The amount of shift of the maximum depends on the range of thicknesses for the sequence l_Σ/λ_{av} considered.

For very thin sequences, the differences in location of the first maximum of the response function is negligible as n varies from 3-7 $[(\Delta l_\Sigma/\lambda_{av}) \approx 0.1]$ (Fig. 2).

For moderately thin sequences (including the interval to the first principal maximum), this difference increases with increasing l_Σ/λ_{av} and for values of l_Σ/λ_{av} which are close to the first principal maximum, the value $\Delta(l_\Sigma/\lambda_{av}) \approx 1$ is reached.

For sequences of moderate thickness, the shift in any particular maximum is somewhat less, with $\Delta(l_\Sigma/\lambda_{av}) \approx 0.2$ as n varies from 3-7.

So, there is a range of values for l_Σ/λ_{av} for which the position of any particular maximum in the response function does not shift along the l_Σ/λ_{av} axis as the number of layers in a sequence is changed. This behavior is important, inasmuch as in many cases it is difficult to distinguish between principal and subsidiary maximums in interpreting experimental curves for $Q_{refl}/Q_{P_1} = \varphi(f)$, and the main criterion used in deciding whether a particular set of data represents a sequence with one or another number of beds is merely a difference in the position of a particular maximum on the frequency axis.

§4. Relationship Between the Wave Forms of Reflected Waves and the Nature of the Periodicity and Number of Beds in a Sequence

Relationship to the Periodicity of Beds in a Sequence

The conditions for interference of waves in a sequence depend on the nature of the periodicity of beds within the sequence. Sequences which have the same form of periodicity of the constituent beds but with different numbers of layers exhibit the same conditions for wave interence. It is obvious that the wave forms of waves reflected from sequences with different periodicities for the constituent beds must be different. Below, we will analyse the forms of waves reflected from sequences with different periodicities, determine the range in values for l_Σ/λ_{av} where the wave forms change or remain constant with various values of S for the sequence, and establish the relation between wave form and the properties of the spectral response function of the sequence, and the corresponding relationship between wave form and the values for the layer parameters in the sequence, and values for l_2/λ_2 and l_1/λ_1.

Recorded Wave Form

Thin Sequences. Thin sequences exhibit a common consistency in the recorded wave forms for reflections independently of the number of layers and the way they repeat in a sequence. For any value of S (with S between 1 and 14), thin sequences ($l_\Sigma/\lambda_{av} = 1.5-2$, with p = 2-4 and n = 3-7) are characterized by a small time difference between the waves reflected from the two surfaces of a layer within the sequence. The properties of the layers comprising a sequence of this type have the following values: $l_2/\lambda_2 \lesssim 0.25$, $l_1/\lambda_1 < 0.4$, and $l_{\Sigma_1}/\lambda_{\Sigma_1} \lesssim 0.5$ (Figs. 3, 4, and 5).

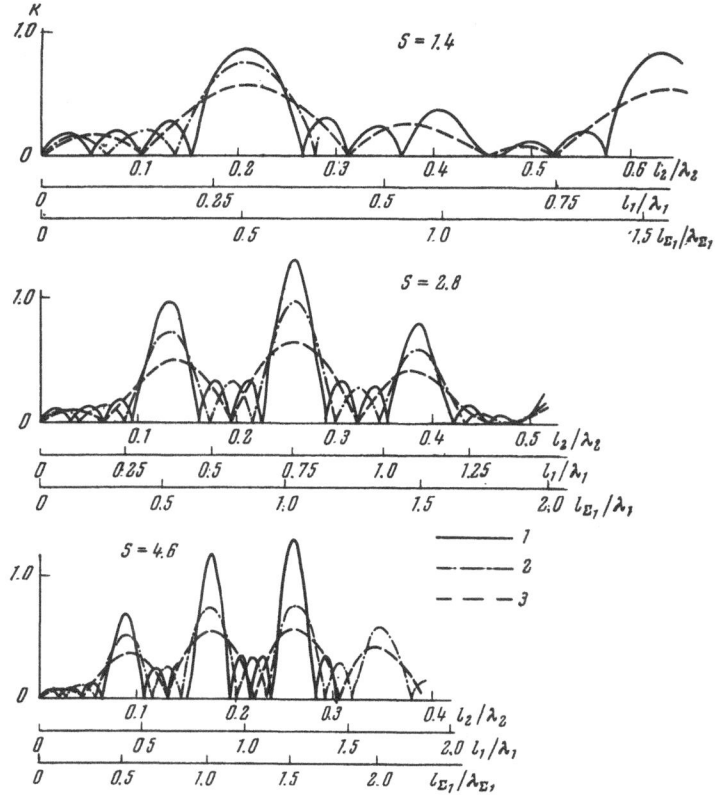

Fig. 3. Relationship of the form of the spectral response
function for wave reflection from a sequence of layers to
the properties of the layers comprising the sequences, with
$V_1/V_2 = 0.7$. 1) $p = 4$; 2) $p = 3$; 3) $p = 2$.

Low-intensity noninterfering reflected waves are associated with the region about the
first subsidiary maximum. The form of the reflected wave closely approximates that of the
theoretical form predicted on the basis of the wave form incident at the layer. The region
about the first minimum is the region of the most complex and least intense wave forms, en-
riched in high-frequency components. The regions about the other subsidiary minimums and
maximums up to the first principal maximum are characterized by complex interference wave-
forms. As one approaches the first principal maximum, the recorded wave form becomes
simpler and has a larger magnitude.

The region about the first maximum, the principal one with respect to energy content, is
a region with large amplitude oscillations, free from interference, and closely approximating
the theoretical wave forms. This region provides the most satisfactory conditions for summing
the waves from the individual beds within the sequence (Figs. 3, 4, and 5). The combined thickness of layers with high and low wave speeds comprises a half wavelength for waves propagat-
ing through them, independently of the value for S; that is, $l_{\Sigma_1}/\lambda_{\Sigma_1} = l_1/\lambda_1 + l_2/\lambda_2 = 0.5$. With
$S = 1$ (sequences with equal τ), the properties of layers in this range are as follows: l_2/λ_2
$= l_1/\lambda_1 = 0.25$; as with the case of a uniform layer, this case corresponds to the optimum summing of waves reflected from the various boundaries in a sequence. The waves are added in
phase. With $1 < S \leq 14$, the properties of the various layers change as follows: $0.03 < l_2/\lambda_2$
< 0.25 and $0.25 < l_1/\lambda_1 < 0.45$; under these conditions, the addition of waves is conducive to the
development of simple but intense reflections.

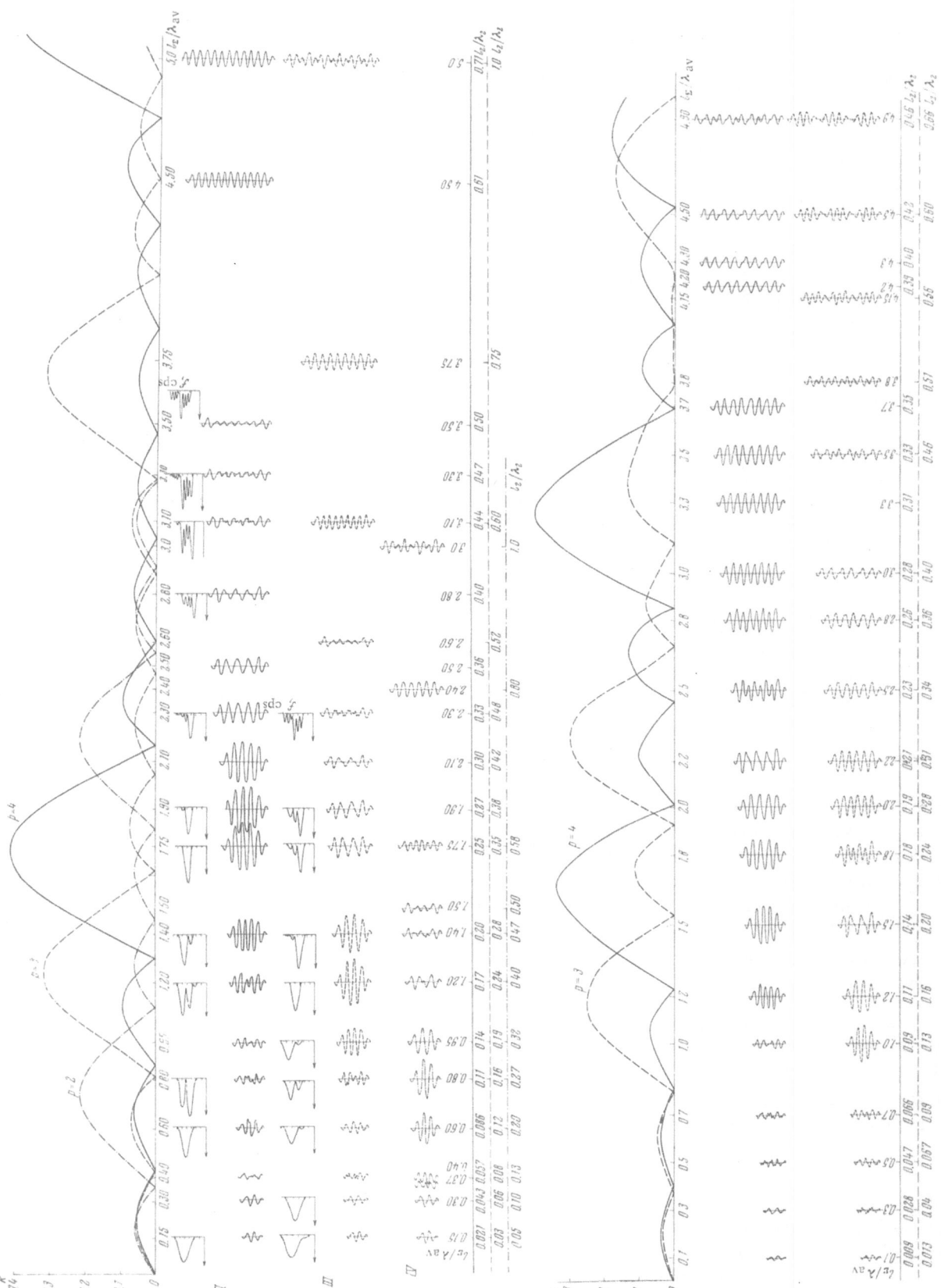

Fig. 4. Synthetic seismograms and spectral response functions for wave reflection from a sequence of layers. a) $S \approx 1$, $l_2/l_1 = 1$; b) $S \approx 2.2$, $l_2/l_1 = 0.5$; I) spectral response function for a sequence; II) synthetic seismogram and spectrums for p = 4; III) the same for p = 3; IV) the same for p = 2.

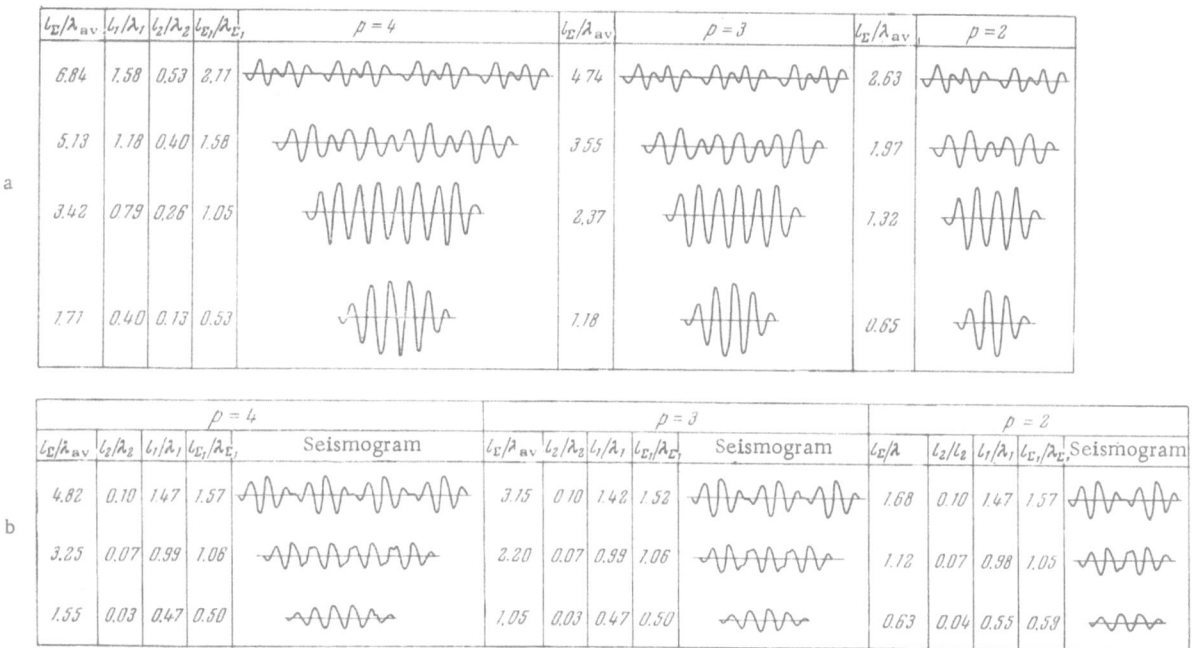

Fig. 5. Synthetic seismograms for waves reflected from a sequence of layers.
a) $S = 3.0$, $l_2/l_1 = 0.37$, $m = 0.9$; b) $S = 14.0$, $l_2/l_1 = 0.08$, $m = 0.9$.

Sequences of Moderate Thickness. With a sequence of moderate thickness,
a complex interference wave is recorded over the regions about the subsidiary maximums between the first and second principal maximums of the spectral response function, independently
of the periodicity of layers within the sequence (Figs. 3, 4, and 5). For values of l_Σ/λ_{av} close
to the region of the second principal maximum of the spectral response function, the character
of these waves depends essentially on the value for S. With $S < 2$, a complex interference wave
of low amplitude is recorded in the region about this maximum, the region with values for l_2/λ_2
close to 0.5. With $S = 1$, the second principal maximum is nearly absent, it becomes subdued,
and l_2/λ_2 is 0.5 in this region (see Figs. 2, 3a). Waves recorded in this region are somewhat
complicated (in fact, only the first and last phases of the oscillation are noted on the record),
and usually would not be recognized above noise level on an actual seismic record. With increasing S (S = 2.5-3, Figs. 2, 3, 4b, and 5a), the nature of the recorded wave form in the region
of the second principal maximum of the spectral response function becomes simple and grows
in amplitude. The largest amplitude waves in this region are observed with S = 3 (Fig. 5a), when
the conditions about the second principle maximum are advantageous for the addition of reflections from the boundaries of layers with high and low wave speeds; in this case, the properties
for the individual layers in the sequence are $l_2/\lambda_2 = 0.25$ and $l_1/\lambda_1 = 0.75$ (Fig. 3). For this case,
the reflected wave form about the region of the second principal maximum is of a comparatively
simple, quasiresonant form. However, with a further increase in S (with S = 1 4, as in Figs. 2
and 5b), the reflected wave begins to contain high-frequency complications, and the recorded
wave form exhibits some interference. This may be explained by the significant time delay
in summing the waves for layers with lower wave speed ($l_1/\lambda_1 \approx 1$) despite the fact that the time delay in the higher speed layers is small and that the value for $l_2/\lambda < 0.25$. Above the region about
the second principal maximum — in the regions about the subsidiary and successive principal
maximums — the recorded form for the reflected waves is complex, showing interference and
a multiple phase structure, independently of the values for S and n. In this region the time delays of waves which interfere from the layers of higher and lower wave speeds are reasonably
large for all values of S.

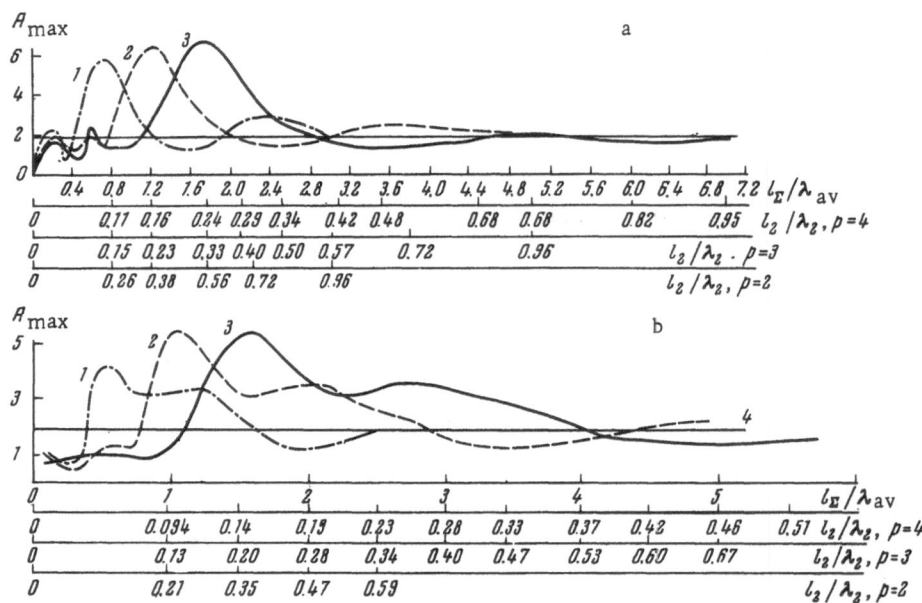

Fig. 6. Nature of the change of maximum amplitude of reflected waves as the thickness of a sequence changes. a) $S \approx 1$, $l_2/l_1 \approx 1$, $V_1/V_2 = 0.9$; b) $S = 2.2$, $l_2/l_1 \approx 0.5$, $V_1/V_2 = 0.9$; 1) $p = 2$; 2) $p = 3$; 3) $p = 4$. The horizontal line represents the value for the reflection factor for waves reflected from a halfspace with the same average values for V and $\bar{\rho}$ as the sequence.

From these considerations, it may be recognized that the form of a reflected wave depends essentially on the time delays of waves in the layers comprising the sequence, such that they interfere with one another. Thus, in any sequence, there will always be a low-intensity interference wave structure if the thickness of the layers with high wave speed is close to a half wavelength ($l_2/\lambda_2 = 0.5$) or if the thickness of the layers with low wave speed is about the same as or larger than a full wavelength ($l_1/\lambda_1 \approx 1$), regardless of the other properties of the layers comprising a sequence (regardless of the value for l_1/λ_1 in the first case, and of l_2/λ_2 in the second case). These values correspond to a sequence of moderate thickness (in the region of the second principal maximum). The strongest and most simple wave forms are observed for a reflection from any sequence in which the combined thickness for a layer with high wave speed and a layer with low wave speed is close to a half wavelength ($l_{\Sigma_1}/\lambda_{\Sigma_1} = 0.5$, $l_2/\lambda_2 \cong 0.25$). The values specified for $l_{\Sigma_1}/\lambda_{\Sigma_1}$ and l_2/λ_2 correspond to thin sequences over the region of the first principal maximum in the spectral response function.

Amplitude Characteristics of Reflected Waves

Figure 6 shows curves for the relationship between the maximum amplitude A of reflected waves and l_{Σ}/λ_{av} for various values of S. The amplitude curves are characterized by a complicated form with well defined extremal points located in the region $l_{\Sigma}/\lambda_{av} < 2$ (with $n \leq 7$) and a minimum in the range of large values for l_{Σ}/λ_{av}.

For thin layers (over the range of values for l_{Σ}/λ_{av} up to the principal maximum of the response function), the curve for $A(l_{\Sigma}/\lambda_{av})$ shows oscillations which are independent of S.

The nature of the amplitude curve for l_{Σ}/λ_{av} in the region of the principal maximum of the response function is diagnostic, and furthermore, it depends on the value for S. For values of S which are not too large (for example, $S \ll 10$), the curve $A(l_{\Sigma}/\lambda_{av})$ is characterized by a maximum point over the region of the first principal maximum of the response function. This is the range of values for l_{Σ}/λ_{av} for which the combined thickness of layers with high and low wave speeds is close to a half wavelength; that is, $l_{\Sigma_1}/\lambda_{\Sigma_1} = 0.5$. The minimum on the curve for $A(l_{\Sigma}/\lambda_{av})$ corresponds to those values for l_{Σ}/λ_{av} for which $l_{\Sigma}/\lambda_2 \approx 0.5$. With increasing S, a maximum is observed on the $A(l_{\Sigma}/\lambda_{av})$ curve for large values of l_{Σ}/λ_{av} in the region of the principal maximum. With $S < 4$ and for values of l_{Σ}/λ_{av} falling close to the principal maximum of the response characteristic, the amplitudes of waves reflected from the sequence are larger by a factor of 2.5-3 than the amplitude of waves reflected from a halfspace having the same average values for V and ρ as the sequence (Fig. 6, curve 4). Thus, the presence of high-amplitude reflected waves, as well as the appearance of interference wave forms, for waves reflected from a sequence of moderate thickness is evidence of the layered structure of the sequence. However, it should be remembered that for very large values of S (the case of very thin layers with high wave speed), the waves reflected from a sequence with $l_{\Sigma}/\lambda_{av} \lesssim 3.5$ are less intense than the waves reflected from a halfspace.

Characteristics of the Reflected Wave Spectrum

The spectrum Q_{refl} of waves reflected from a thin layer or sequence of thin layers may be considered to be the product of a specified spectrum for the incident waves and the spectral response function for the layer or sequence, the spectral response function for wave transmission through the overlying rocks and the spectral attenuation function for the overlying rocks [6, 7]. For various values of total sequence thickness; that is, for various values of l_{Σ}/λ_{av}, the spectral response function for wave reflection from a sequence of layers, as well as that for a uniform layer [6] is "scaled" along the frequency axis; that is, a specific frequency scale for the spectral response function for each l_{Σ}/λ_{av}. Consequently, the form of the spectral response function will differ over any specified frequency range for different values of l_{Σ}/λ_{av}; that is, different numbers of maximum and minimum points may be noted over the same range of frequencies. Moreover, with various values for l_{Σ}/λ_{av} and forms for the reflected wave spectrum (with a fixed spectrum for the incident waves) the function will be characterized by different numbers of maximum and minimum points, and the appropriate properties of the reflected wave spectrum will be observed over a reasonably narrow frequency band, if it is assumed that the spectrum of the incident waves is given over a limited frequency range.

Thin Sequences (Fig. 4). For thin sequences in the range of values for l_{Σ}/λ_{av} up to the principal maximum, the nature of variation of the reflected-wave spectrum will be practically the same for sequences having different bed periodicities. The range of values for l_{Σ}/λ_{av} characterized by a common form of spectrum changes only by a minor amount.

Regardless of the value for S, the reflected-wave spectrum reproduces the incident-wave spectrum for values of l_{Σ}/λ_{av} corresponding to the regions about the first subsidiary maximum and the principal maximum in the spectral response function; the region about the first subsidiary maximum has values $l_{\Sigma}/\lambda_{av} < 0.2$, and the region about the first principal maximum has values $l_{\Sigma}/\lambda_{av} = 0.6-0.8$ for $p = 2$ and $l_{\Sigma}/\lambda_{av} \approx 1.75$ for $p = 4$. The relative width of the reflected wave spectrum changes only insignificantly in comparison with the width of the incident-wave spectrum. Wave interference is observed in the regions about the other subsidiary maximum points; auxiliary maximum points appear in the reflected-wave spectrum which are not present in the incident-wave spectrum. The number of auxiliary maximums for a given l_{Σ}/λ_{av} depends on the width of the incident-wave spectrum. For spectrums with a relative width $\delta f \approx 50\%$ (see Fig. 4), the number of auxiliary maximums in the region of the principal maximum is commonly no more than one ($p = 4$, $l_{\Sigma}/\lambda_{av} < 1.75$).

Sequences of Intermediate Thickness. Regardless of the value for S, the reflected-wave spectrum also exhibits interference and auxiliary maximum points, for values of

l_Σ/λ_{av} lying between the first and second principal maximums in the spectral response function (for example, when $p=4$, $l_\Sigma/\lambda_{av} = 1.75$-$3.0$). The number is usually 2 or 3.

The character of the spectrum for values of l_Σ/λ_{av} corresponding to the second principal maximum does depend on the value for S. With $2 > S \geq 1$, spectrums for these values of l_Σ/λ_{av} are complicated and exhibit interference. With $S \geq 2$, with a reasonably narrow spectrum for the incident waves ($\delta f \approx 50\%$), the reflected-wave spectrum over the region of the second principal maximum of the response function repeats the incident wave spectrum (the range $l_2/\lambda_2 < 0.5$).

Over the range of values for l_Σ/λ_{av} above the second principal maximum, the reflected-wave spectrum always shows interference for any value of S, consisting of several maximums with the primary ones being associated with large-amplitude components and the subsidiary ones with smaller-amplitude components (see Fig. 4 with $p = 4$, $l_\Sigma/\lambda_{av} > 3$).

Thus, changes in the periodicity of layers in a sequence have a primary effect on the form of waves reflected from a sequence only in the case of a sequence of moderate thickness, and then not for all values of l_Σ/λ_{av}.

Relationship to the Number of Layers in a Sequence

In this section, we have studied the relation of the form of a reflected wave to changes in the number of layers in a sequence with a specified total thickness, considering sequences with various specific values for l_Σ/λ_{av}. These relationships have been analyzed for sequences with various periodicities to the layers.

As has been mentioned previously, the effect of the number of layers in a sequence on the form of reflected waves does not depend on the nature of the periodicity S of the layers in the sequence. We have indicated above that the characteristic wave forms, spectrums and amplitudes change uniquely with the variation of the number of layers in a sequence. A change in the number of layers in a sequence leads to change in the range of values for l_Σ/λ_{av} for which a particular wave form is noted, for a specified periodicity. The extent of regions of l_Σ/λ_{av} with interfering or simple wave forms or spectrums increases with an increase in the number of layers in a sequence. With increasing n, a peculiar change in spectrum and amplitude is observed for larger values of l_Σ/λ_{av}. In the case of very thin layers ($l_\Sigma/\lambda_{av} < 0.4$), a change in the number of layers does not affect the recorded wave forms.

§5. The Possibility of Using the Form of Reflected Waves for Identifying and Tracing the Structure of Periodic Sequences of Layers with Varying Thickness

Using the solution of the direct problem developed in the preceding paragraphs for a periodic sequence, we may draw conclusions concerning the use of the reflected wave form for identifying a sequence and determining the number of layers and nature of periodicity in the sequence. We will analyze with this intention, all of the wave form characteristics for reflected waves.

Identification of Sequences of Variable Thickness

The computations described above allow us to establish how reflected wave forms will change as the total thickness of a sequence gradually changes: a) because of a change in the number of beds comprising the sequence; b) because of a decrease in the thickness of the high-speed layers; or c) because of a decrease in the thickness of all the layers comprising the sequence. Studies of cases (b) and (c) have shown that the properties of the recorded wave forms vary in the same manner. In both cases, a repetitious sequence of changes in the re-

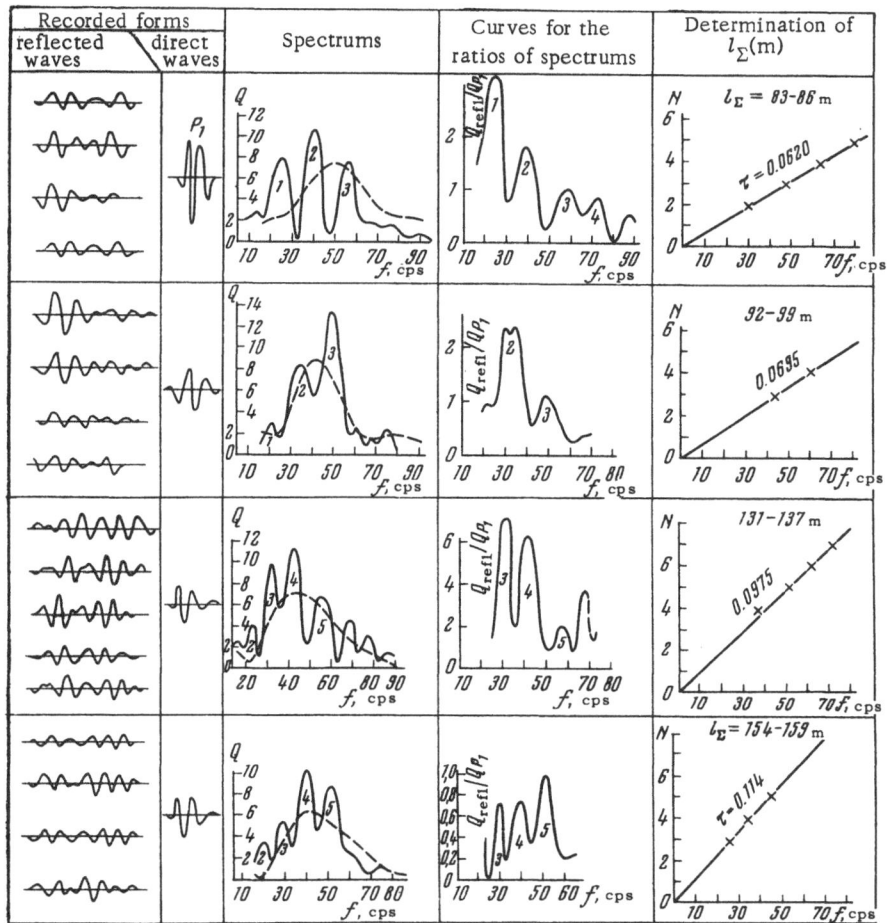

Fig. 7. Example of the experimental recognition of and determination
of the thickness of a sequence of clastic rocks with varying thickness.
The resultant character of the reflected wave forms is shown to the
left; the numbers indicate the maximums in the spectrums.

corded wave forms is observed as the sequence thickness changes; in the case of thin sequences, simple wave forms and spectrums change to complicated interference forms as the
thickness of the sequence increases, and then return to a simple form (see §4). The length
of a recorded wave form increases as the thickness increases, characteristically varying
the amplitudes. With increasing thickness, a diagnostic change in the extremals of the reflected-wave spectrum is observed; the number of extremal points in a given frequency band
increases, the spacing Δf between adjacent extremals decreases, and the maximum is shifted
towards lower frequencies (Fig. 7). These properties of the spectrum are more reliable than
changes in duration or recorded wave forms for reflected waves. Moreover, the ranges of
l_Σ/λ_{av} for which changes in the thickness l_Σ have practically no effect on the recorded wave
form (Fig. 4) may be identified while correlation of spectrums permits determination of the
direction of growth in thickness of a sequence (Fig. 4). Thus, the spectral properties of waves
are most sensitive to changes in the thickness of a sequence of layers.

Determining the Nature of the Periodicity of Layers in a Sequence

Consideration of the effect of changes in the periodicity of layers in a sequence on reflected wave forms has shown that, in most cases, there is no unique solution to the inverse

TABLE 1 *

PR	Shot point	Thickness given by seismic data, l_Σ, in m	Thickness found in bore hole, m	Percentage difference
	1280	83—86	70	+ (18—23)
	1820	92—99	92	+ (0—7.7)
3	3380	131—137	145	— (5.5—9.6)
	4680	154—159	163	— (2.4—5.5)
	2590	65	~50	~+30
5	3030	70—72	~53	~+32—36
	3250	64—66	~55	+16—18
	5500	93—95	75—80	+ (16—25)
7	5760	88—92	90—95	— (2—8)
	6380	95—98	105—110	— (2.0—14)

* This table lists results obtained from data recorded with type SS-60-51D seismic equipment by a party from the Krashodar Branch of the Scientific Research Institute for Geophysics. The Institute and the Academy of Sciences were conducting field surveys at the same time.

problem for a periodic sequence of layers. It is practically impossible to establish the nature of the periodicity of layers in a thin sequence from reflected wave forms.

As may be deduced from §4, the most favorable situation for determining the periodicity of layers in a sequence of moderate thickness is with values for l_Σ/λ_{av} close to the region about the second principal maximum of the spectral response function. Such determinations are possible if the approximate thickness of the sequence in the survey area is known. Finding the value for S applicable to a sequence being studied experimentally leads to a method for selecting the appropriate theoretical model for the sequence. The most satisfactory approach may be to select sequences for which S varies from 1 to 2.

As a consequence, there is only a slight possibility of determining the nature of the periodicity of layers in a sequence by using the diagnostic features of the reflected wave form. Chances are better if a method proposed by S. A. Kats is used.*

Determining the Number of Layers in a Sequence

A determination of the number of layers in a sequence may be accomplished for thin sequences over the range of values for l_Σ/λ_{av} about the principal maximum of the response function. One may establish the number of layers in a sequence under consideration by establishing the range of values for l_Σ/λ_{av} (values for $l_\Sigma/\bar{V} = \frac{1}{2}\tau$) from the form of the reflected wave for which the nature of the record is the same as for the region about the first principal maximum in the response function (§4). Determinations of the number of layers in a sequence made over other ranges of values in l_Σ/λ_{av} are not satisfactory.

§6. Determining the Total Thickness of Periodic and Some Nonperiodic Thin Sequences

Periodic Sequences

The possibility of approximating a periodic sequence of layers (all with essentially the same τ) as a uniform layer with properties close to the averaged properties of the sequence

---

*S. A. Kats. Several methods for increasing resolution in seismic exploration. Candidate's Dissertation, Fondi Inst. Fiz. Zemli, Akad. Nauk SSSR, 1965.

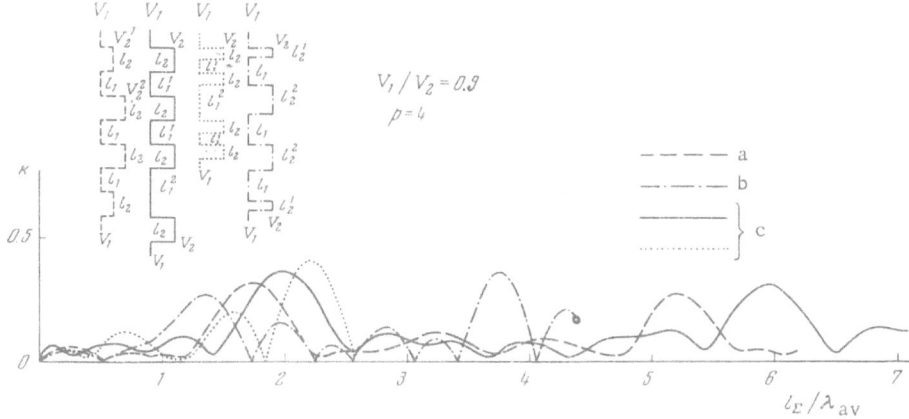

Fig. 8. Spectral response functions for several nonperiodic sequences.
a) First type; b) second type; c) third type.

was first pointed out in reference [1]. It was pointed out that the extremal points of the spectral response function of the sequence (in comparison with the same function for a uniform layer) might be used to establish the equivalent thickness of the uniform layer satisfactorily. Reference [2] points out that the total thickness of a sequence may be determined, with some error, from the interval between two subsidiary minimums in the spectral response function plotted. In both cases, the sequence thickness is determined from the interval Δf between subsidiary minimums, using the average wave speed of all the beds [1], or using the same average wave speed in a unit of the sequence, that is, a pair of layers with high and low wave speeds [2].

Here, we consider the possibility of determining the thickness of a periodic sequence over various ranges in values for l_{Σ}/λ_{av} for changes in the nature of periodicity in the sequence.

1. For sequences with equal τ ($S = 1$), the maximum error in determining the total thickness l_{Σ} is about 0.1 $(\Delta l_{\Sigma}/\lambda_{av})$, which amounts to 2-15 m in sequences which have wave speeds ranging from 2500-4000 m/sec (with $f_{pr} = 40$-50 cps).

2. Changes in the periodicity in a sequence (an increase in S) lead to increased errors in determining the total thickness of the sequence; moreover, errors are larger with sequences having smaller n.

For changes in S by an order of magnitude, as for example from 1.4 to 14, the error for a many-layer sequence increases only slightly (for $n = 7$, $\Delta(l_{\Sigma}/\lambda_{av})$ increases by 0.03), while for sequences with only a few beds, the error may be doubled (for $n = 3$, $\Delta(l_{\Sigma}/\lambda_{av})$ increases to nearly 0.12). The total error in determining l_{Σ} in these cases is as follows: $n = 7$, $S = 1.4$, $\Delta(l_{\Sigma}/\lambda_{av}) = 0.09$; $S = 14$, $\Delta(l_{\Sigma}/\lambda_{av}) = 0.12$; $n = 3$, $S = 1.4$, $\Delta(l_{\Sigma}/\lambda_{av}) = 0.16$; $S = 14$, $\Delta(l_{\Sigma}/\lambda_{av}) = 0.28$.

Thus, only in the case of sequences with a large number of constituent beds is the thickness determined with reasonable accuracy, regardless of the nature of the periodicity of beds in the sequence.

3. Regardless of the nature of the periodicity, the thickness of a sequence can be determined within these error limits only for those ranges of values for l_{Σ}/λ_{av} which cover a band of frequencies in the reflected-wave spectrum having at least two subsidiary minimums in the spectral response function; that is, the same condition as for a uniform layer.

Examples of the determination of the thickness of the first Maikop horizon in the Krasnodar Belt from seismic data are given in Table 1 and Fig. 7, along with a comparison with geological information. The spectrums of the waves reflected from these layers have properties

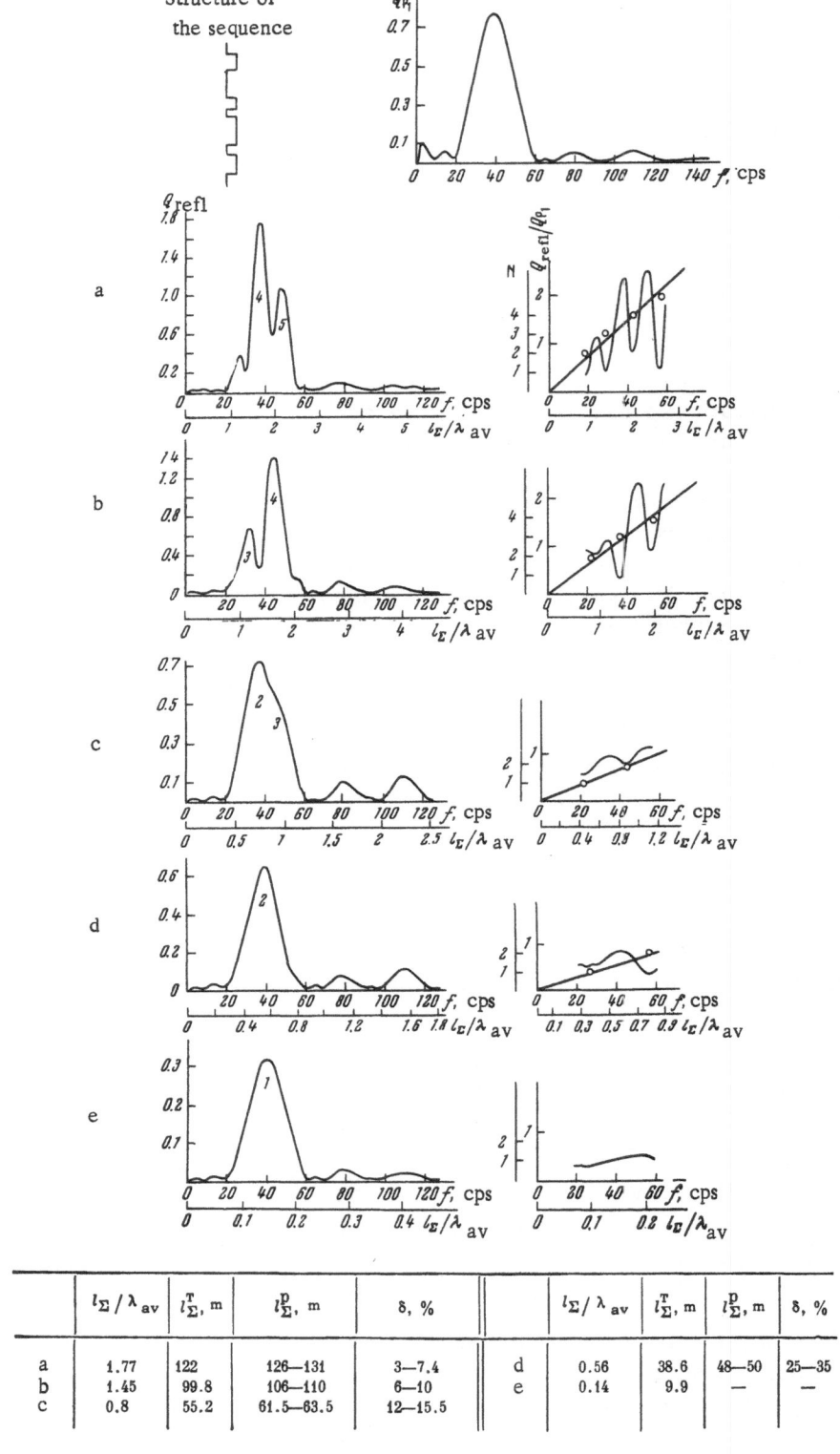

	l_Σ/λ_{av}	l_Σ^T, m	l_Σ^p, m	δ, %		l_Σ/λ_{av}	l_Σ^T, m	l_Σ^p, m	δ, %
a	1.77	122	126—131	3—7.4	d	0.56	38.6	48—50	25—35
b	1.45	99.8	106—110	6—10	e	0.14	9.9	—	—
c	0.8	55.2	61.5—63.5	12—15.5					

Fig. 9. Spectrums for waves reflected from nonperiodic sequences, and examples of the determination of total thicknesses for various values of l_Σ/λ_{av}, for $V_1/V_2 = 0.9$.

similar to those for the spectrum for a uniform layer. In determining sequence thicknesses, the spectrums were analyzed statistically, using auxiliary curves for $\varphi(f) = Q_{refl}/Q_P$ obtained over a wide frequency range (with various filters); the average wave speed in the sequence was taken to be 2700-2800 m/sec in calculating l_Σ.*

The discrepancies in determining l_Σ from seismic data and from drilling amount to 2.5-36%, which reflects most of all the lack of reliable information about $\bar{V}$ in the Maikop beds.

Nonperiodic Sequences

Several examples of spectral response functions for sequences with n = 7 were computed on a high-speed digital computer. Four types of nonperiodic sequences were selected: a) periodicity in thicknesses was preserved, but not in wave speeds (Fig. 8a); 2) periodicity in wave speeds was preserved but not in bed thicknesses (Fig. 8b); and 3) the layers were practically periodic with respect to thickness (Fig. 8c). Consideration of Fig. 8 indicates that in all cases, the spectral response functions have diagnostic principal and subsidiary extremals. For sequences with complete or partial periodicity with respect to thickness, and a lack of periodicity with respect to wave speeds, the spectral response functions are more or less similar for the range of values for $l_\Sigma/\lambda_{av} \leq 2$; that is, they have the same number of subsidiary maximums below the principal maximum, and the periodicities of the maximums are about the same. The response function for case (2) differs markedly from the others even in this range of values for l_Σ/λ_{av}. On the basis of these data, we may conclude that apparently the lack of even a part of the periodicity with respect to thickness has a greater effect on the form of the spectral response function than a lack of periodicity in wave speeds.

Determinations of thicknesses based on the subsidiary maximums in the response function have shown that with nonperiodic sequences of types (1) or (3) over the range $0.6 < l_\Sigma/\lambda_{av} \leq 2$, the error in total thickness is no more than 20%. For case (2), the error in determining l_Σ was more than 25%. Thus, it is apparent that the spectral response function may be used to determine the total thickness of some types of nonperiodic sequences (Fig. 9).

It will be necessary to obtain data on nonperiodic clastic bed sequences using acoustic logging data before final conclusions may be reached.

Conclusions

1. We have presented the results of a computational study of wave forms reflected from periodic sequences, with various characteristics for the layers comprising the sequences.

2. Regardless of the number of layers, periodic sequences follow consistent rules in the mechanics of forming reflected waves. The corresponding response functions for these sequences also show general characteristics. The relation between the form of a reflected wave and the spectral response function for a periodic sequence of layers has been established.

3. The waves for any type of sequence will always be complicated when the thickness of the layers with high wave speed is close to a wavelength ($l_1/\lambda_1 \approx 1$). The strongest waves are observed when the combined thickness of a layer with high wave speed and a layer with low wave speed is close to a half wavelength ($l_{\Sigma_1}/\lambda_{\Sigma_1} = 0.5$). The relationship between the form of the reflected waves and the nature of the periodicity of layers within a sequence has been studied. The effect of the periodicity of beds in a sequence on the wave form does not depend on the number of layers in a sequence. Apparently, not every change in the periodicity of layers

---

*Described in reports in 1964-1965 on the topic "Simple physical models of real media and the study of wave propagation in them," Vol. II, Fondi Inst. Fiz. Zemli, Akad. Nauk SSSR, 1965.

is seen in the reflected wave form. Changes in the periodicity of beds in a sequence affect the wave form in different manners, depending on the ratio of the total thickness of a sequence to a wavelength l_Σ/λ_{av}.

For thin sequences, the reflected wave form does not depend on the nature of periodicity of the beds in a sequence; a change in the nature of periodicity of the beds does not affect the form of the waves reflected from sequences even of moderate thickness.

4. The relationship between the form of the reflected waves and the number of beds in a sequence has been studied. A change in the number of layers in a sequence leads to a change in the range of values for l_Σ/λ_{av} for which one or another specific wave characteristic is noted, for waves reflected either from thin sequences or sequences of moderate thickness.

5. Conclusions have been drawn about the possibility of determining the structure of a sequence on the basis of reflected wave forms:

a) determinations of layer periodicities are nonunique in most cases; and

b) a selective method may be used in determining the number of layers in a sequence; the main criterion is the range of values observed for l_Σ/λ_{av} over which diagnostic features in the record occur about the principal maximum in the response function.

6. The possibility of determining the total thickness of a periodic sequence for various values of S and n and various ranges of values for l_Σ/λ_{av} has been considered. For a sequence with many beds, the thickness may be determined with an error $\Delta(l_\Sigma/\lambda_{av}) \leq 0.1$, regardless of the value for S. The range over which the thickness may be determined with an error $\Delta(l_\Sigma/\lambda_{av}) \leq 0.15$ was determined.

7. The possibility of determining the total thickness of several types of nonperiodic sequences with n = 7 was indicated with various examples. The errors in determining l_Σ are no more than 20% with $l_\Sigma/\lambda'_{av} < 2$.

LITERATURE CITED

1. N. G. Mikhailova, B. S. Pariiskii, and M. V. Saks, Frequency response of sequences of layers, Izv. Akad. Nauk SSSR, Fiz. Zemli, No. 1 (1966).
2. V. S. Isaev, Results of the application of the theory of grouping in the study of periodic sequences of layers, Prikl. Geofiz., Vol. 49 (1967) [English translation: Exploration Geophysics, Vol. 49, Consultants Bureau, New York, 1969].
3. S. P. Starodubrovskaya, Physical basis for the use of wave forms of longitudinal reflected waves for tracing layers with varying thickness, Izv. Akad. Nauk SSSR, Fiz. Zemli, No. 12 (1964).
4. I. S. Berzon, A. M. Epinat'eva, G. N. Pariiskaya, and S. P. Starodubrovskaya, Wave Forms of Seismic Waves in Actual Media, Izd. Akad. Nauk SSSR, 1962.
5. S. P. Starodubrovskaya and G. N. Pariiskaya, Use of wave form characteristics of reflected waves for discovering layers of variable thickness, Razvedochnaya Geofizika, No. 2 (1964).

THE EFFECT OF THE DENSITY RATIO
AT A BOUNDARY ON THE FORM OF THE
SPECTRAL RESPONSE OF A THIN LAYER

T.G. Ivanova

In recent years, theoretical calculations of the amplitudes of waves of various types, of the spectral response of layers, and so on for specific experimental conditions have been used frequently in interpreting seismic data. The principal desire has been the determination of the seismic properties of the section: the thickness of individual layers and the propagation speeds for elastic waves within these layers. Usually, the densities of the rocks are not considered in such calculations. The reason for this is that information about densities is normally not available in areas where seismic exploration is being done. As a consequence, the density ratio $\rho_k/\rho_{k\pm1}$ at a boundary is usually assumed to be unity in the calculations. In fact, though, it is well known that rocks which differ in age or lithology also differ in density. Actually, the range of relative change in the density of various media is considerably smaller than the range of relative change in wave speeds for the same rocks.

The question of the effect of the contrast in densities between layers in contact with each other on the energy of the waves transmitted through a boundary or through a thin layer has been considered in a number of papers [1-3, and others]. Results are presented [1] showing the effect of a contrast in densities $\rho_k/\rho_{k\pm1}$ in liquid layers and in a solid layer immersed in a liquid medium on the energy of a longitudinal wave propagating through the layer. However, as was pointed out in reference [3], the application of these results to the case of a solid medium is not always permissible.

The results of computations of the coefficients for reflection and refraction for solid thin-layered media with different values for density in the contacting media have been given [2, 3].

To supplement these earlier studies, I have considered the effect of the ratio $\rho_k/\rho_{k\pm1}$ on the spectral response of a solid thin layer situated in a homogeneous elastic space in this paper.

Using previously developed expressions [1], I have calculated the spectral response of a layer for the transmission of longitudinal PP and dilatational PS waves. The calculations were carried out for a layer of low wave speed for the following properties of the medium: $V_{P_2}/V_{P_1} = 0.5$, $V_{S_1}/V_{P_1} = V_{S_2}/V_{P_2} = 0.6$, and $\rho_2/\rho_1 = 0.7, 0.8$, and 1.0, with V_{P_1} and V_{S_1} being the speeds for longitudinal and transverse waves in the medium surrounding the layer, V_{P_2} and V_{S_2} being the speeds for longitudinal and transverse waves in the layer, and ρ being the density. The angle of incidence i at the lower boundary of the layer was taken equal to 30° or 60°. Spectrums were calculated over the interval $0 \le l/\lambda_2 \le 1.0$, where $\lambda_2 = V_{P_2}/f$ (f is the frequency).

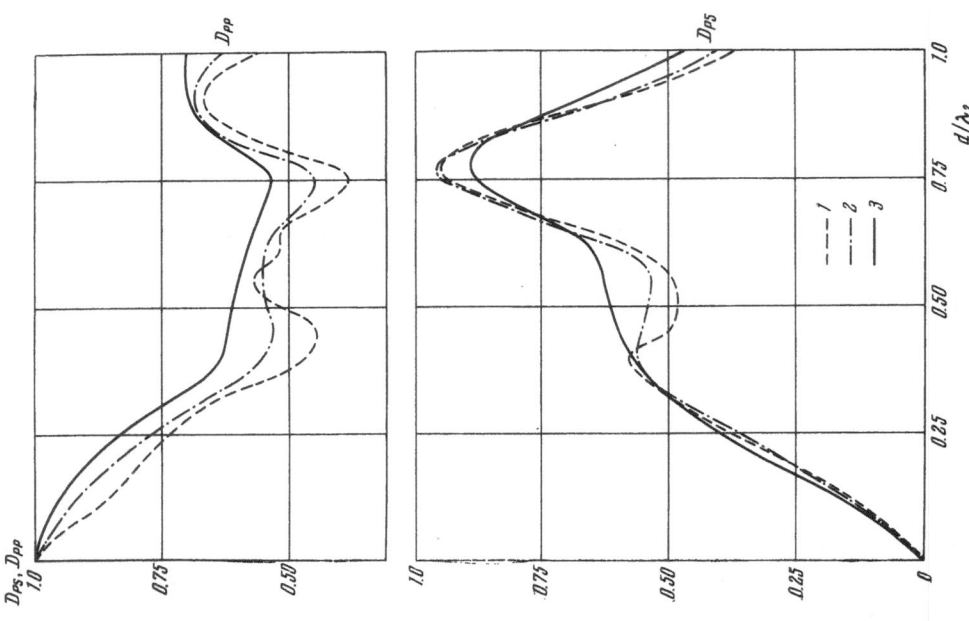

Fig. 2. The spectral response of a layer in terms of the transmission of longitudinal PP and dilatational PS waves for various values of the ratio ρ_2/ρ_1. 1) 0.7; 2) 0.8; 3) 1.0; angle of incidence, 60°.

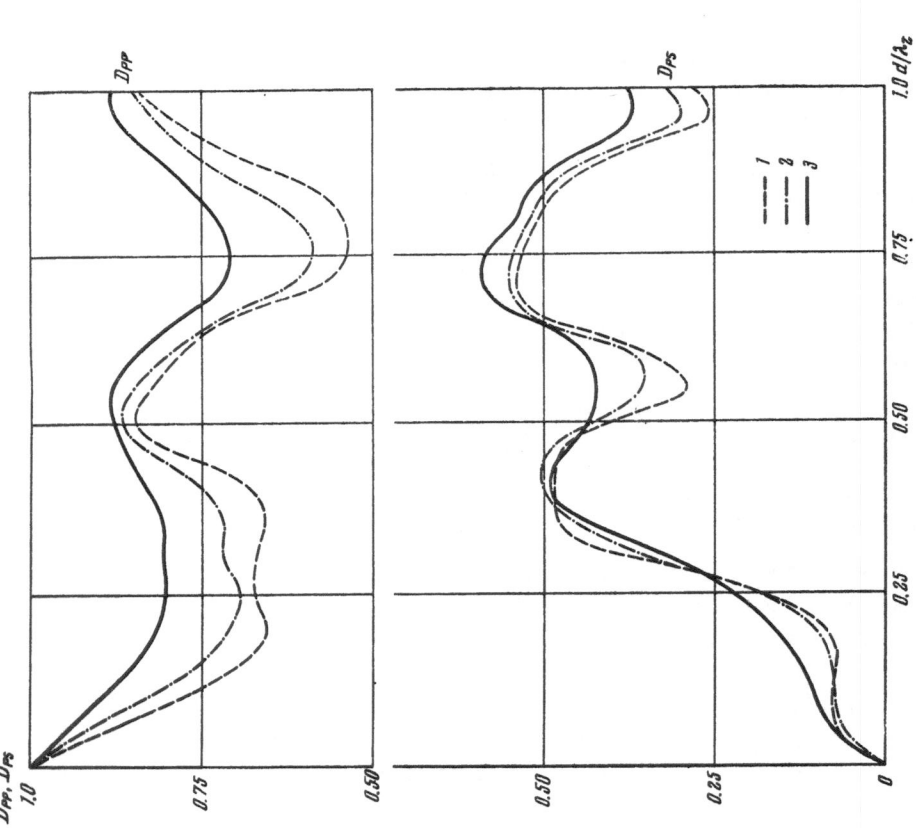

Fig. 1. The spectral response of a layer in terms of the transmission of longitudinal PP and dilatational PS waves for various values of the ratio ρ_2/ρ_1. 1) 0.7; 2) 0.8; 3) 1.0; angle of incidence, 30°.

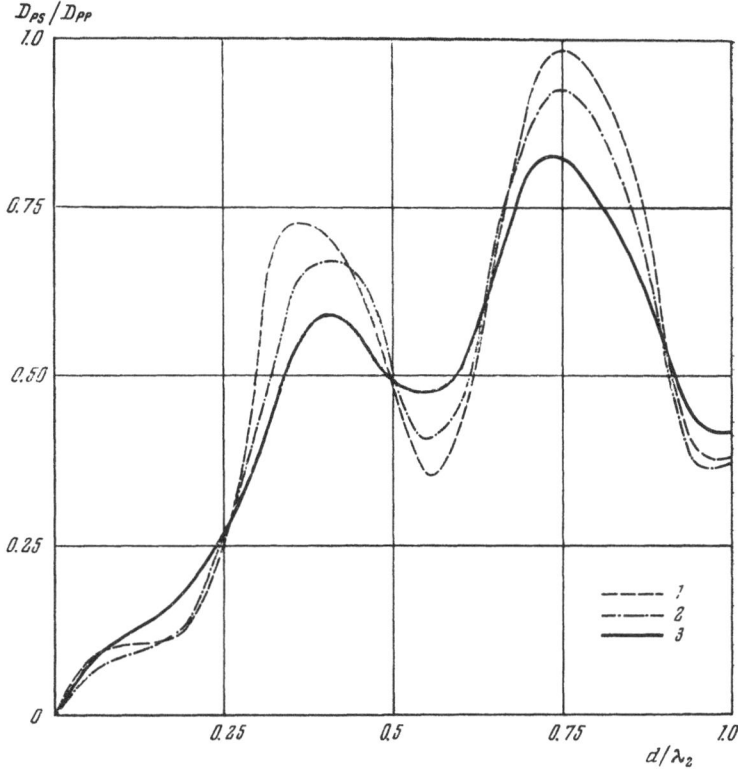

Fig. 3. The ratio of spectral characteristics of a layer for dilatational PS and longitudinal PP waves for various values of ρ_2/ρ_1. 1) 0.7; 2) 0.8; 3) 1.0. Angle of incidence, 30°.

The results of these computations are shown in Figs. 1 and 2 in the form of the relationship of the coefficients for wave transmission D_{PP} and D_{PS} to the function l/λ_2. It is apparent from these curves that for the case $i = 30°$, the curves for D_{PP} as well as for D_{PS} have a uniform character for various values of ρ_2/ρ_1. The greatest spread in the absolute values calculated for the transmission coefficients is found about the minimums on the curves. There, the spread in the values for the transmission coefficients for density contrasts of 0.7 and 1.0 may reach 25-30%. The differences between these same curves about their maximums are insignificant.

We will now examine the case with the larger angle of incidence at the layer (Fig. 2). In addition to the same sort of differences about the extremals which were noted in the previous case, at $i = 60°$, as the density ratio ρ_2/ρ_1 is decreased from unity, subsidiary extremal points appear on the curves for D_{PP} and D_{PS}. The presence of these subsidiary maximum points in the spectral response for a layer, even though of relatively small amplitude, may lead to the appearance of new complexities in the spectrums for PP or PS waves propagating through the layer. That is, if the maximum of the spectrum of the incident wave falls about values l/λ_2 which correspond to the subsidiary maximum for the spectral response of the layer, while the principal maximum of the spectrum is situated on the right slope of the spectral response, the spectrum of the transmitted waves may even have two maximum points in extreme cases. As a result, it would be difficult to explain the complicated form of the spectrum for transmitted PP and PS waves which might be observed experimentally, using theoretical calculations only for the case $\rho_2/\rho_1 = 1.0$.

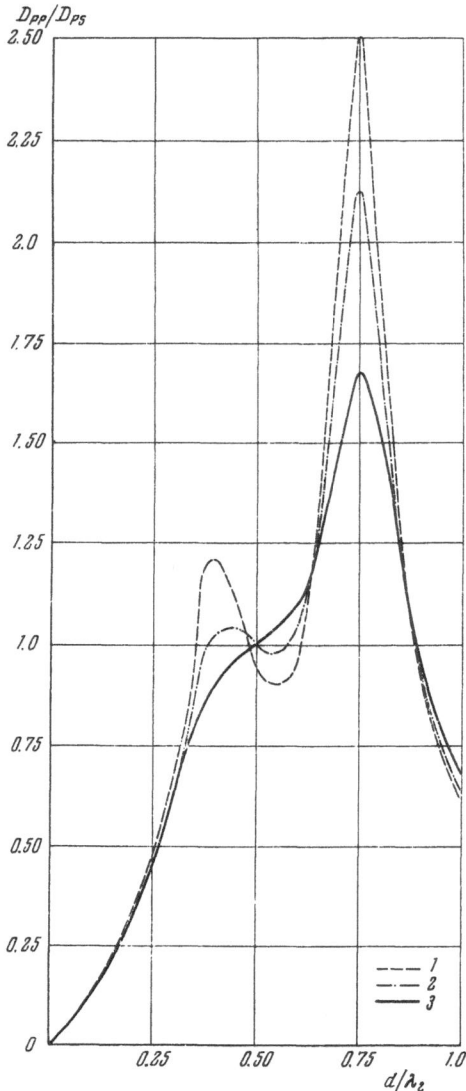

Fig. 4. The ratio of spectral characteristics of a layer for dilatational PS and longitudinal PP waves for various values of ρ_2/ρ_1. 1) 0.7; 2) 0.8; 3) 1.0. Angle of incidence, 60°.

Determinations of the ratios of various wave types transmitted through the same layer may be incorrect over some ranges of values for l/λ_2 if ρ_2/ρ_1 is taken as unity. Curves for the ratio D_{PS}/D_{PP} as a function of l/λ_2, based on these same calculations, are shown in Figs. 3 and 4. It is apparent from these curves that the spread in values for the relative intensities of PS and PP waves in the region about the maximum points on the spectral response curves may amount to 20% for i = 30° and to 35-45% for i = 60° when ρ_2/ρ_1 is varied from 0.7 to 1.0.

Similar calculations for weak contrasts in propagation speeds, with $V_{P_2}/V_{P_1} = 0.9$, have shown that differences in the transmission coefficients for D_{PP} and D_{PS} waves may be ignored in this case. Weaker velocity and density contrasts may be ignored.

Thus, the effect of a density difference at a boundary between layers with a strong contrast in wave speeds, on the spectral response of a layer is marked. Therefore, it is not satisfactory to assume that the ratio of densities is unity in calculations. I have indicated the desirability of including consideration of density changes at boundaries in the reduction of seismic data. If information on the density of the section is not available, it is necessary at least, to evaluate the possible errors which may occur as a result of ignoring density changes.

LITERATURE CITED

1. I. S. Berzon and A. M. Epinat'eva, On seismic screening, Izv. Akad. Nauk SSSR, Ser. Geogr. i Geofiz., No. 6 (1950).
2. L. I. Molinovskaya, Methods for constructing synthetic seismograms, in: Aspects of the Dynamic Theory of the Propagation of Seismic Waves, No. 1, Gostoptekhizdat, 1957.
3. Yu. I. Vasil'ev, Comparison of coefficients for the reflection and refraction of elastic waves at the boundary of two solid or two liquid media, Izv. Akad. Nauk SSSR, Ser. Geofiz., No. 9 (1959).
4. G. S. Pod'yapol'skii, Coefficients of refraction and reflection of elastic waves by a layer, Izv. Akad. Nauk SSSR, Ser. Geofiz., No. 4 (1961).

PART THREE

METHODS FOR AND RESULTS FROM THE STUDY OF WAVE PROPAGATION IN REAL MEDIA

METHODS FOR AVERAGING THE
FORMS OF RECORDED WAVES

V.V. Kun

It is necessary to improve the precision with which the wave form of a seismic wave is determined in the solution of a number of seismic exploration problems. One such problem is the choice of an appropriate representation of the medium by comparing computed wave forms with those obtained experimentally. Coincidence of computed wave forms with observed wave forms may be one of the primary criteria in selecting a representative model of the medium. This requires increased accuracy in recording the wave form in the field. More precision is also required for the use of recorded wave forms for the solution of the inverse problem.

One method for increasing the precision in observing wave forms is statistical averaging. The problem of obtaining an averaged (or stacked) wave form has been developed in a paper by V. F. Pisarenko and T. G. Rautian [1], in which the effect of such factors as the location of an earthquake and siting of a seismic station on the period of some of the waves recorded from earthquakes was evaluated. This paper showed that the effect of these factors is important, and that as a result, averaging should be considered in evaluating precision; methods were developed for evaluating precision, considering the effects of various factors.

In the present paper, I have considered some aspects of methods for obtaining an average of the wave forms recorded in seismic exploration operations. The study was carried out for a high-intensity reflected wave, obtained at normal incidence in the northern part of the Krasnodar Belt. The first part of this paper is an evaluation of the noninterdependence of the data used in stacking, the accuracy of the results of stacking, and the stability of the stacked data in relation to the time base over which stacking is performed. The second part of the paper explores the question of the choice of an optimum method for recording and analysis, leading to precise results. Very simple wave forms have been selected as examples for the consideration of wave properties. In studying stacking, the question of the accuracy with which spectrums may be computed on a computing machine may also arise.

The methods used in this paper are quite similar to those used in reference [1]. However, it must be realized that the material is still of interest because the seismic prospecting problem is quite different from those studied in earthquake seismology.

Brief Description of Waves and the Methods for Stacking Wave Forms

High-intensity reflected waves are recognized in the northern areas of the Krasnodar Belt at times $t_0 \approx 0.3$ sec (a wave designated as $P_{0.3}$) and $t_0 \approx 1.3$ sec (a wave designated as $P_{1.3}$). Examples of the $P_{0.3}$ and $P_{1.3}$ events are shown in Fig. 1. With shot depths of 15 to 25 m, the $P_{0.3}$ and $P_{1.3}$ events are partially distorted by ghost arrivals, representing the first reflection

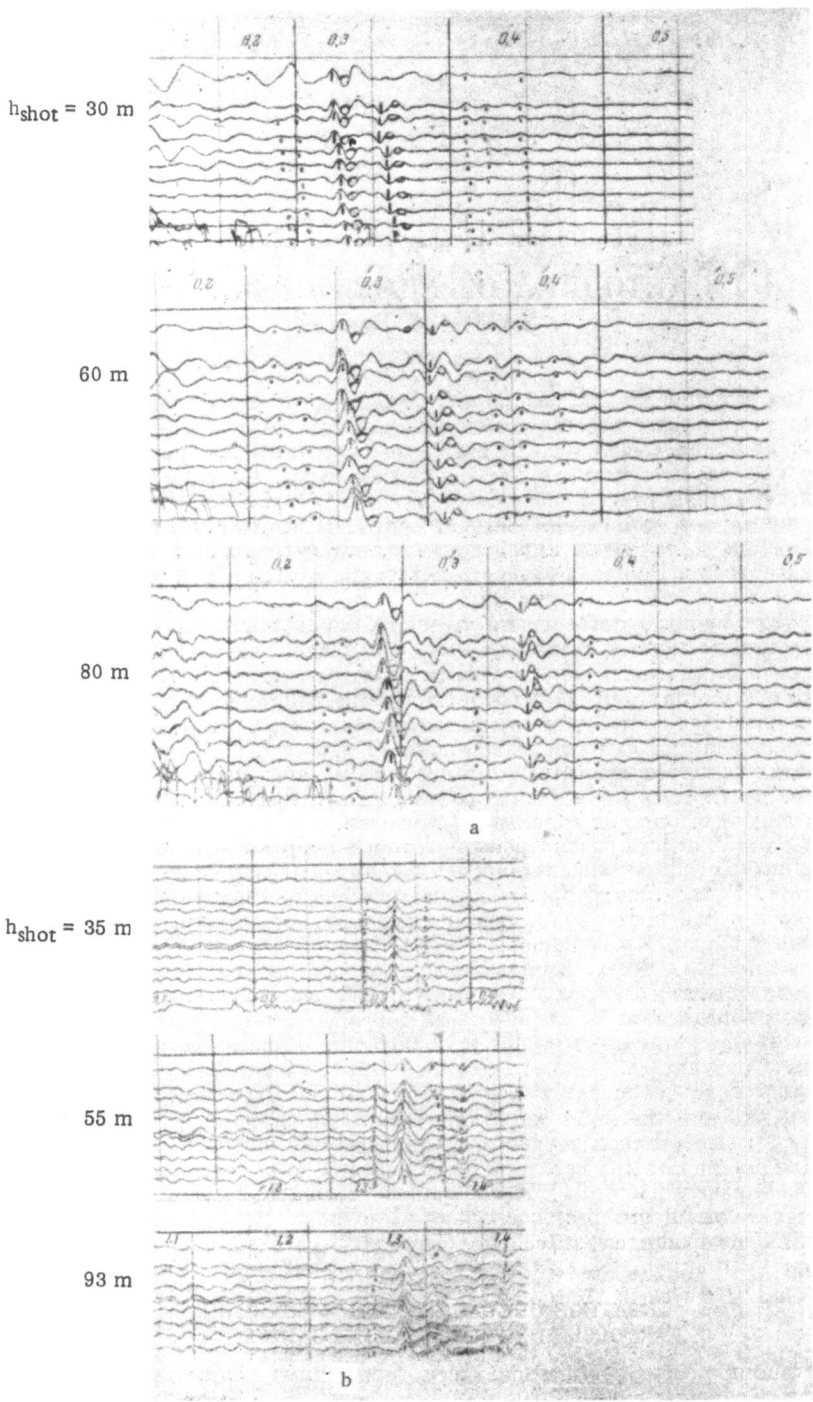

Fig. 1. Seismic records with the events $P_{0.3}$ (a) and $P_{1.3}$ (b), obtained with various shot depths h. In addition to the events $P_{0.3}$ and $P_{1.3}$, the ghosts $P_{0.3}^g$ and $P_{1.3}^g$ may be recognized.

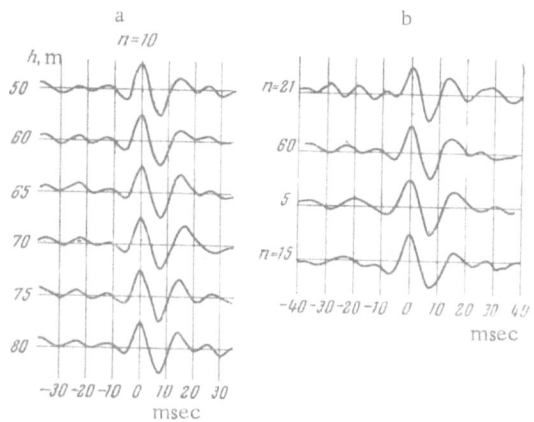

Fig. 2. Stacking of the $P_{0.3}$ event obtained at one site for different shot depths (a) and at different sites for the same shot depth (b). n is the number of stacks; $\Delta x \approx 0\text{-}100$ m. (The stacked records in Fig. 2a were obtained from the seismic records in Fig. 1a).

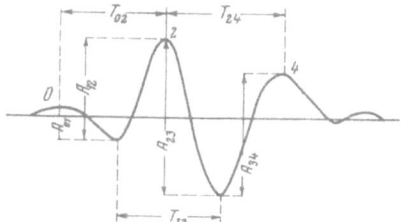

Fig. 3. Definition of characteristics describing the $P_{0.3}$ event.

from the base of the low-velocity layer. At greater shot depths (> 45 m), the ghost arrivals are well separated from the primaries (see Fig. 1).

In stacking records, data were used with slight changes in the shooting conditions and geophone plants, as well as in the separation between the shot hole and the geophone spread.

Records were made along longitudinal profiles with lengths of 40-150 m, with the spacing between geophones being 5-10 m. A standard charge weight of 0.3 kg was used in shooting. Shots were fired at 1-5 m intervals over depths of 50-100 m.

Waves were recorded with broadband filter settings; mixing, delay, and gain control were not used.

Stacking of the wave forms was accomplished by taking the arithmetic average of values for like wave forms. The stacking process was carried out as follows.

1. A record selected for stacking was reduced to a specified gain (normalized), and then it was digitized at a predetermined interval Δt of 0.001 sec or 0.0005 sec. The position of the primary maximum swing was taken to be zero time (Fig. 2). Normalization was accomplished by multiplying all of the ordinates by a factor k which reduced the amplitude of the record at t = 0 to a fixed value.

2. At each time t (t = 0.001, 0.002, ...; −0.001, −0.002, ... sec) the digitized values were averaged.

Examples of averaged records for the $P_{0.3}$ event are shown in Fig. 2.

The record now may be considered to be a combination of a repetitive signal and random oscillations. These random oscillations reflect the combined effects of geophone plant conditions, shot hole conditions, low-velocity noise events, noise vibrations, clutter noise from the shot, microseisms, and so on. If the effects of each of these factors on the records are independent; that is, if they will not be added systematically in the stacking process, the recorded data may be considered to be uncorrelated. Then, in averaging a large number of events, the effect of all of these factors tends to zero. An evaluation of the accuracy of the stacking process with regard to various recording parameters was carried out using standard mathematical tests for the noninterdependence of random variables [2]. If the effect of several, or even one of the factors is detectable, it must be considered in evaluating precision [1, 2].

In addition to the factors enumerated above, there may be an effect from a systematic change in the signal wave form at different points along a profile. The cause of such changes may be variations in the reflection factor at boundaries as a result of a change in angle of incidence, or attenuation.

TABLE 1

Wave characteristic	$P_{0.3}$		$P^{(1)}_{1.3}$		$P^{(2)}_{1.3}$		γ_ξ			γ_η			γ_ε			n_{ef}		
	F_ξ	F_η	F_ξ	F_η	F_ξ	F_η	$P_{0.3}$	$P^{(1)}_{1.3}$	$P^{(2)}_{1.3}$	$P_{0.3}$	$P^{(1)}_{1.3}$	$P^{(2)}_{1.3}$	$P_{0.3}$	$P^{(1)}_{1.3}$	$P^{(2)}_{1.3}$	$P_{0.3}$	$P^{(1)}_{1.3}$	$P^{(2)}_{1.3}$
T_{02}	1.22	4.04	0.69	7.59	3.62	24.6	0.09	0.004	0.017	0.255	0.43	0.74	0.655	0.57	0.24	16.01	16.40	10.13
T_{13}	3.34	4.52					0.26	0.050		0.22	0.48		0.52			13.9	14.2	
T_{24}	3.78	4.95	1.80	1.01			0.28	0.17		0.22	0.33		0.50	0.48		13.7	15.6	
A_{01}/A_{23}			3.38	6.85										0.50				
A_{12}/A_{23}	1.21	3.50	4.32	2.70	4.20	6.59	0.10	0.30	0.19	0.23	0.13	0.35	0.67	0.58	0.45	17.0	19.0	13.3
A_{34}/A_{23}	6.50	3.82	12.85	4.84	12.08	2.37	0.45	0.54	0.43	0.12	0.11	0.26	0.43	0.35	0.31	13.9	13.8	11.0
Average value	3.21	4.18	4.63	4.60	6.65	11.15	0.236	0.213	0.213	0.209	0.296	0.450	0.55	0.497	0.334	14.9	15.8	11.5

TABLE 2

Wave characteristic	γ_ξ	γ_η	γ_ε	n_{ef}
T_{02}	0.037	0.475	0.486	14.2
T_{13}	0.260	0.220	0.520	14.5
T_{24}	0.165	0.350	0.490	14.0
$\dfrac{A_{12}}{A_{23}}$	0.197	0.237	0.567	16.4
$\dfrac{A_{34}}{A_{23}}$	0.473	0.163	0.363	12.9
$\overline{T}_{(02,\,13,\,24)}$	0.116	0.392	0.495	—
$\overline{A}_{12,\,34}$	0.312	0.219	0.470	—
Average value	0.222	0.298	0.481	14.0

TABLE 3

Wave characteristic	$P_{0.3}$		$P^{(1)}_{1.3}$		$P^{(2)}_{1.3}$	
	$\dfrac{\sigma_{single}}{\bar{x}}$	$\dfrac{\bar{\sigma}}{\bar{x}}$	$\dfrac{\sigma_0}{\bar{x}}$	$\dfrac{\bar{\sigma}}{\bar{x}}$	$\dfrac{\sigma_0}{\bar{x}}$	$\dfrac{\bar{\sigma}}{\bar{x}}$
T_{02}	13.6	3.5	4.5	1.1	6.5	2.1
T_{13}	4.9	1.2	—	—	—	—
T_{24}	8.5	2.4	9.7	2.8	—	—
A_{12}/A_{23}	14.8	3.7	11.4	2.7	16.4	4.7
A_{34}/A_{23}	15.1	4.2	12.8	3.6	14.5	4.6
Average value	11.4	3.0	9.6	2.5	12.5	3.8

$\dfrac{\sigma_0}{\bar{x}}$, %; $\dfrac{\bar{\sigma}}{\bar{x}}$, %

Inasmuch as the question of the precision of the stacking process may arise, we will now consider the effects of various factors on stacking and evaluate the noninterdependence of data.

Geophone Plant Conditions, Shot Hole Conditions, and Random Noise

Factors which may effect any single digitized point in a random manner fall into three groups: geophone plant conditions, shot hole conditions, and random noise. Under geophone plant conditions, we include the combined effects of the contact between the geophone and the soil, nonhomogeneity in the upper part of the section, and nonidentical response of geophones. Shot hole conditions determine the spectrum of the excitation. All other factors are considered to be random noise (wave noise, microseisms, vibration noise, shot hole clutter).

We have evaluated the effect of geophone plant conditions, shot hole conditions, and random noise on the following characteristics describing the $P_{0.3}$ and $P_{1.3}$ events: on the periods T_{02}, T_{13}, and T_{24}, and on the ratios of amplitudes of phases, $A_{01}/A_{23} : A_{12}/A_{23} : A_{34}/A_{23}$. The definitions of these parameters are given in Fig. 3.

Using the various procedures for analysis of dispersion given in reference [1], as well as techniques which we have developed, we have evaluated the effects of shot hole conditions, geophone plant conditions and random noise on the results of stacking; we have evaluated the precision of stacking. In so doing, we have considered the following characteristics:

1. F_ξ and F_η are ratios characterizing the effect of geophone plant conditions (ξ) and shot hole conditions (η). They are determined using a Fisher distribution

$$F_\xi = \frac{S_1^2}{L-1} \cdot \frac{N-L-M+1}{\Delta S}, \qquad F_\eta = \frac{S_2^2}{M-1} \cdot \frac{N-L-M+1}{\Delta S},$$
$$\Delta S = S^2 - S_1^2 - S_2^2, \tag{1}$$

where S is a general estimate of dispersion, ΔS is an estimate of residual dispersion, S_1^2 and S_2^2 are estimates of dispersions according to parameters, L and M are the numbers of geophones and shot points, respectively, and N is the number of observations.

Comparison of values for F_ξ and F_η with tables would permit determining the significance of the effects from geophone plant conditions and shot hole conditions.

2. γ_ξ, γ_η, and γ_ε are coefficients having the meaning of dispersion weights, reflecting the influence of the various factors σ_ξ^2, σ_η^2, and σ_ε^2 on the total dispersion σ^2 for a specific variable:

$$\gamma_\xi = \frac{\sigma_\xi^2}{\sigma^2}, \quad \gamma_\eta = \frac{\sigma_\eta^2}{\sigma^2}, \quad \gamma_\varepsilon = \frac{\sigma_\varepsilon^2}{\sigma^2}, \quad \gamma_\xi + \gamma_\eta + \gamma_\varepsilon = 1$$

(ε characterizes the effect of random factors). The coefficient γ permits a quantitative evaluation of the effect of each factor and will be used here in evaluating precision.

3. n_{ef} is a number indicating the equivalent degrees of freedom in a series of N observations, in which the effect of factors η and ξ appear in addition to a random error, ε.

The value for n_{ef} determined from (1) is

$$\frac{1}{n_{ef}} = \frac{\gamma_\xi}{L} + \frac{\gamma_\eta}{M} + \frac{\gamma_\varepsilon}{ML}. \tag{2}$$

Values have been computed for F, γ, and n_{ef} for wave characteristics for the $P_{0.3}$ event recorded at well No. 45 in the Staro-Minsk field and the $P_{1.3}$ event recorded at wells No. 39 and No. 52 in the Leningrad field. For the $P_{0.3}$ event, L = 10 and M = 6, while for the $P_{1.3}$ event, L = 10 and M = 6-8.

Table 1 is a list of values for F, γ, and n_{ef} for various record characteristics $\bar{x}$, and averages determined for each characteristic.

Table 2 is a list of average values for γ_ξ, γ_η, γ_ε, and n_{ef} computed for each record characteristic from stacked data for the $P_{0.3}$, $P_{1.3}^{(1)}$ and $P_{1.3}^{(2)}$ events.*

We may draw the following conclusions on the basis of the data in Tables 1 and 2.

1. The effects of geophone plant conditions and shot hole conditions are significant (γ_ξ and γ_η are comparable in size with γ_ε), and they must be considered in evaluating the precision of stacked records. The significance of these factors is indicated by the fact that F_ξ and F_η are larger than the tabulated values by a factor of 2.1, for a 5% level of confidence [2].

2. The effect of geophone plant conditions on the average is nearly equal to the effect of shot hole conditions, and is half as large as the effect of random factors: $\gamma_\xi = 0.222$, $\gamma_\eta = 0.298$, and $\gamma_\varepsilon = 0.481$ (see the last column in Table 2, where values for γ_ξ, γ_η, and γ_ε calculated for five wave characteristics are listed.

3. The effect of random noise (ε) is nearly half of all the effects considered ($\gamma_\varepsilon \approx 0.50$). This result was obtained quite consistently for both of the waves considered.

4. The effects of the factors ξ and η on the various wave characteristics are different:

a) On the average, the wave period T is more dependent on shot hole conditions than on geophone plant conditions; that is, $\gamma_\eta = 0.392$ while $\gamma_\xi = 0.116$ (Table 2);

b) more than any of the other characteristics, the ratio of amplitudes A_{34}/A_{23} depends on the geophone plant conditions; for both waves the effect of the factor ξ on A_{34}/A_{23} is larger than that of the factor η and even larger than the effect of random noise; the reason for this behavior is not known.

5. Because of the influence of the shot hole conditions and the geophone plant conditions which contribute systematic errors to the data, the effective degrees of freedom η_{ef} is much less than the actual number of observations N. The average value for η_{ef} is determined with high accuracy for each of the waves, because it does not differ much for the various wave characteristics.

The value for n_{ef} is less than that for N by a factor of 4-6, and 13-16 for N = 60-80.

Accuracy of the Stacked Characteristics

The range in uncertainty for the average value of a parameter $\bar{x}_0$, at the 5% confidence level, is [1, 2]

$$\bar{x} - \frac{\lambda_\alpha S}{\sqrt{n_{ef}-1}} \leqslant x_0 \leqslant \bar{x} + \frac{\lambda_\alpha S}{\sqrt{n_{ef}-1}}. \tag{3}$$

Here, $\bar{x}$ is the average value for a parameter determined from stacking, and S is the dispersion for single determinations of x:

$$S^2 = \frac{1}{N-1} \sum_{i=1}^{N} (x_i - \bar{x})^2,$$

and λ_α is obtained from tables for normal distributions for a level α.

---

* The event $P_{1.3}$ has been designated as $P_{1.3}^{(1)}$ at well No. 39 and as $P_{1.3}^{(2)}$ at well No. 52.

Table 3 lists values for the relative standard error for single determinations of the parameter x and its stacked value $\bar{x}$, in percent (with N = 60-80): $\sigma_{single}/\bar{x}$ and $\bar{\sigma}/\bar{x}$. The value for $\bar{\sigma}$ is determined from the formula

$$\bar{\sigma} = \frac{\sigma_{single}}{\sqrt{n_{ef}-1}}.$$

It is apparent from the table that the precision in determining various record characteristics is not uniform. It is one and one-half to two times higher for the wave period than for amplitude ratios. The relative errors for single determinations of the parameters under consideration are no more than 15-16%. The average value for these errors, computed for all waves and all characteristics, is 11%.

The relative error in determining stacked characteristics ($\bar{\bar{\sigma}}/\bar{x}$, percent) for the n_{ef} obtained lies within the limits 1.1-4.7%, and is 3.1% on the average. Thus, stacking such large numbers of experimental data increases the precision by a factor of four on the average. Such a precision is completely adequate for a 5% level of confidence. Furthermore, substituting a value of 2.13 for λ_α for a 5% confidence level [2] in Eq. (3) and with $n_{ef} = 14$, we have

$$\bar{x} - 0.57S \leqslant x_0 \leqslant \bar{x} + 0.57S.$$

Thus, the ratio $\bar{\sigma}/\bar{x}$ need not be more than 0.5S (that is, $0.5\sigma_{single}$ in our symbology in Table 3). It is apparent from Table 3 that this condition is met, inasmuch as $\bar{\sigma}$ is less than σ_{single} for a factor of 3-4.

Uniformity of Data in Relation to Increasing the Offset Distance in Stacking

We will now consider what the effect would be on the scatter of the observed data with changes in the wave form caused by changes in the angle of arrival at the Earth's surface; that is, what the combined effects of changes in Δx would be on the uniformity of the observed data (Δx is the offset distance from the shot point to the geophone).

Uniformity of records with respect to Δx was tested in the following manner. A group of sixty records of the $P_{0.3}$ event was divided into two subgroups: 1) records obtained with Δx of 0-50 m; and 2) records obtained with Δx of 60-100 m.*

Using the average values for the record characteristics in terms of periods and amplitude ratios for each of the subgroups, we tested the hypothesis that the two were independent subgroups (1 and 2) of a single general population (that is, we tested the randomness or significance of the differences between the average values for the parameters). In this we used a Student's t-test [2]:

$$t = \frac{\bar{x}_1 - \bar{x}_2}{S}\sqrt{\frac{n_1 n_2}{n_1 + n_2}}. \tag{4}$$

Here, $\bar{x}_1$ and $\bar{x}_2$ are the average values for each of the parameters in the subgroups 1 and 2, and S is the total dispersion,

$$S^2 = \frac{(n_1 - 1)S_1^2 + (n_2 - 1)S_2^2}{(n_1 - 1)(n_2 - 1)}; \tag{5}$$

---

*The emergence angle for the $P_{0.3}$ wave varied 0-10° for this range in Δx.

TABLE 4

	T_{02}, msec	T_{13}, msec	T_{24}, msec	$\dfrac{A_{12}}{A_{23}}$	$\dfrac{A_{34}}{A_{23}}$
$\bar{x}_1$	12.05	12.94	14.78	0.626	0.846
$\bar{x}_2$	12.49	12.86	14.51	0.679	0.818
n_{ef}	8.8	8.6	8.8	9.2	10.2
t	0.312	0.332	0.256	0.0690	0.025

S_1 and S_2 are the dispersions of the subgroups, and n_1 and n_2 are the number of statistically independent data in each subgroup (that is, n_{ef}).

The values of n_{1ef} and n_{2ef} were determined from Eq. (2) for N = 30 by substituting the values for γ_ξ, γ_η, and γ_ε for the $P_{0.3}$ wave from Table 3.

The resultant determinations of $\bar{x}_1$, $\bar{x}_2$, t, and n_{ef} are listed in Table 4. It is apparent from a comparison of these values with the tabulated values for t [2] that the differences between $\bar{x}_1$ and $\bar{x}_2$ are not significant (t < 2), and may be explained by random scatter of the values $\bar{x}_1$ and $\bar{x}_2$. Therefore, it may be concluded that for Δx of 0-100 m, there is no systematic change in the wave form as a function of Δx, and it is reasonable to do stacking over this range in off-set distances. This result was obtained for the $P_{0.3}$ event, but it should be even more valid for the $P_{1.3}$ event, for which the emergence angle changes even less.

It is apparent from Table 2 that t ≪ 2. Because of this, we may conclude that the offset distances used in stacking may be increased significantly.

Recording and Selecting Records for Stacking

We will now consider what methods for recording and analyzing records should be used in order to obtain high precision in stacking with a minimum amount of recording and analysis. We will use Eq. (2) for this analysis with the values for γ_ξ, γ_η, and γ_ε from Table 2 substituted in it. Such an analysis was done in reference [1]. We have obtained essentially the same results as were reported in reference [1].

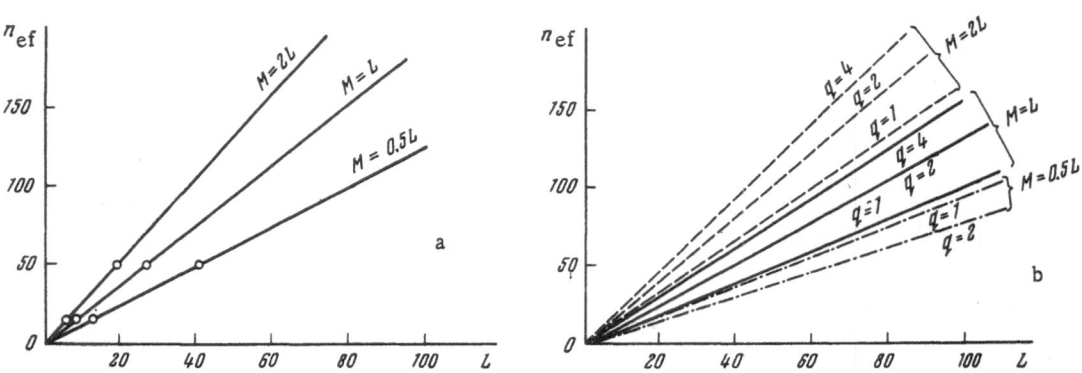

Fig. 4. Curves showing the dependence of n_{ef} on the number of geophones L.
a) For stacking using all traces on the seismic records for all shot points; b) for stacking using q traces from each seismic record. M is the number of shots.

TABLE 5

$\bar{x}$	$\bar{x}_0$	$\bar{\sigma}$				$\bar{\sigma}/\bar{x}_0$, %				$\dfrac{\bar{x}-\bar{x}_0}{\bar{x}_0}$, %	$\dfrac{\bar{x}-\bar{x}_0}{\bar{x}_0}$, %
		a	b	c	d	a	b	c	d		
T_{02}	12.33	0.865	0.613	0.638	0.593	7.02	4.97	5.17	4.81	2.5	11.1
T_{13}	12.91	0.321	0.270	0.111	0.0520	2.49	2.09	0.86	0.40	0.8	4.2
T_{24}	14.62	0.618	0.698	0.392	0.282	4.23	4.77	2.68	1.93	1.4	8.5
A_{12}/A_{23}	0.653	0.0473	0.0359	0.0294	0.0101	7.24	5.50	4.50	1.54	2.8	2.3
A_{34}/A_{23}	0.831	0.0510	0.0858	0.0395	0.0170	6.14	10.32	4.75	2.04	4.7	6.7
Average $\dfrac{\bar{\sigma}}{\bar{x}_0}$, %						5.42	5.53	3.59	2.14	2.4	6.5

Inasmuch as we are concerned with all of the strong waves recorded in given regions, we substitute values for γ_ξ, γ_η, and γ_ε from the last column of Table 2 into Eq. (2), where the average values for γ_ξ, γ_η, and γ_ε for all parameters and all waves are listed:

$$\frac{1}{n_{ef}} = \frac{0.22}{L} + \frac{0.30}{M} + \frac{0.48}{LM}. \qquad (6)$$

It is obvious from Eq. (6) that n_{ef} does not increase without limit as L (or M) increases, but has a limit:

$$\lim_{L\to\infty} n_{ef} = \frac{M}{\gamma_\eta} = \frac{M}{0.30}; \qquad \lim_{M\to\infty} n_{ef} = \frac{L}{\gamma_\xi} = \frac{L}{0.22}.$$

With M = 1, and L → ∞, $n_{ef} \leq 3.3$, and for L = 1 and M → ∞, $n_{ef} \leq 4.5$. It is obvious that the precision of the stacked parameters will not increase significantly if a large number of geophone spreads are used with a single shot or if a single geophone spread is used with a large number of shots. It is necessary to increase both M and L.

Figure 4a shows curves for the dependence of values of n_{ef} on the number of geophones L, with M parametric, computed from Eq. (6). They allow us to determine the value for n_{ef}, and so also, the precision in determining the average wave characteristic for specified values of M and L. Using these, we may also determine the number M necessary to provide a specified precision for a given number L, and so on. It is apparent from Eq. (6) and Fig. 4a that a large volume of data must be used, with stacking of a large number of traces in order to get a significant increase in precision.

It was indicated in reference [1] that a higher precision may be obtained with fewer computations if large numbers are used for M and L, with stacking not being done for all traces from each record. The minimum number of stacked traces is required when only one trace from each record is used in stacking (q = 1)* and the number of shots equals the number of geophone spreads (M = L). In this case, the data may be considered to be independent, and $\eta_{ef} = M = L$.

The use of several traces from each shot (q = 2, 3, 4) also reduces significantly the volume of computations from that needed when all traces (q = L) are stacked. Figure 4b shows

*q is the number of traces used in stacking from each seismic record, that is, from each shot.

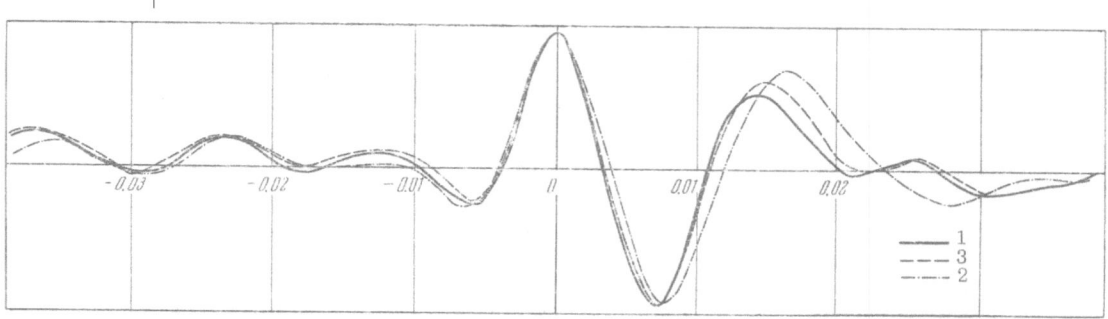

Fig. 5. Comparison of three stacked records for the $P_{0.3}$ event, with stacking being done in various ways: 1) from the entire data set (N = 60, L = 10, M = 6); 2) type-b stacking (h_{shot} = 70 m, L = 10); 3) type-d stacking (N = 12).

curves for $n_{ef} = f(L)$ with parameter q, calculated, in accordance with reference [1], from the expression

$$\frac{1}{n_{ef}} = \frac{0.22}{L} + \frac{0.30}{M} + \frac{0.48}{qM}. \tag{7}$$

It is apparent from Fig. 4 that when q is increased, a point of diminishing returns is quickly reached with respect to improved precision of the stacked characteristics. For example, with M = L = 20, the precision is increased by a factor of 4.5 with q = 1, by a factor of 5.5 with q = 4, and by a factor of 6 when q = L = 20. The difference in the number of stacked records with q = 1 and with q = L is very significant (N = 20 for q = 1 and N = 400 for q = L).

Examples of Stacking with Various Choices. We will show that the standard deviation (σ) in determining $\bar{x}$ is larger when $\bar{x}$ is determined for choices where M or L = 1 than when $\bar{x}$ is determined from independent data, that is, for small q.

Table 5 is a list of the absolute and relative values for the standard deviation of values for the characteristic $\bar{x}$ for the $P_{0.3}$ wave for four different approaches:

a) L = 1, M = 6, N = 6;
b) M = 1, L = 10, N = 10;
c) M = L = 6, q = 1, N = 6;
d) M = 6, L = 10, q ≈ 1.5, N = 12.

The average value of $\bar{x}_0$ taken for the whole group of data with L = 10, M = 6, and N = 60 is taken as the true value.

Values for $\bar{x}$ were computed as the average of six relations for each procedure. The value $\bar{\sigma}$ characterizes the standard deviation for the departures of values for the parameter x from the true value (that is, the departure of $\bar{x}$ from $\bar{x}_0$).

It is apparent from a comparison of Tables 4 and 5 that an average of x with methods "a" or "b" has twice the precision of a determination based on individual values ($\bar{\sigma}/\bar{x}_0 = 11\%$ in Table 4, but 5.5% in Table 5). The averages obtained with methods "c" and "d" have higher precisions by factors of 3 and 4 ($\bar{\sigma}/\bar{x}_0 = 3.5$ and 2.1), so that they are better.

An example of a record obtained by stacking with technique "b" is shown in Fig. 2a. It is apparent from Fig. 2a that the stacked record may be quite significantly different than the "true" record compiled from the larger (more general) group of data, for certain values of h_{shot} (the second record in Fig. 2b). Thus, for example, the record obtained with h_{shot} = 70 m is a markedly lower-frequency record than either those obtained for other shot depths or the "true" record shown in Fig. 2b. Two traces are shown plotted in an expanded coordinate system in Fig. 5; trace 1 was obtained from the entire set (N = 60) of data, and trace 2 was obtained by stacking 10 of these 60 records (with a constant shot depth of 70 m). The differences between the characteristics of trace 2 and those of trace 1 are listed in the last column of Table 5. Apparently, the differences average about 6.5%. A third trace, obtained by using the second type of stacking procedure, is also shown on Fig. 5. It is obviously much closer to trace 1 than to trace 2.

The differences between the characteristics for trace 1 and 3 amount to 2.4% on the average (next to last column in Table 5).

Conclusions

1. Stacking of strong reflections has been accomplished and the effectiveness of the technique demonstrated.

2. A significant effect from geophone plant conditions and shot hole conditions on the wave form has been established; quantitative estimates of the effects of various factors were made. It was shown that the sum of the dispersion weights due to geophone and shot hole conditions was approximately equal to the dispersion weight contributed by random errors.

3. Curves were compiled which permit an evaluation of the precision of the stacked record characteristics as a function of the number of shots, the number of geophones, and the number of traces stacked. These curves may be used to plan a survey system which will provide a required precision for the stacked results. It was pointed out that it is necessary to consider these factors when the precision of stacking is being considered. The curves are valid for strong reflections under the conditions prevailing in the northern part of the Krasnodar Belt.

4. It was pointed out that the most logical and economical way to stack records is one which makes use of statistically independent data. In order to obtain such independence in field data, one should use a multiplicity of shots and geophones. The use of one shot point with many geophone locations or many shots with a single geophone location is a less satisfactory approach.

LITERATURE CITED

1. V. F. Pisarenko and T. G. Rautian, The effect of station factors and noise on the accuracy of determining seismic parameters. In: Analysis of Seismic Data by Computer. Computational Seismology, Vol. 1 (revised), "Nauka" (1966).
2. A. K. Mitropol'skii, Techniques of Statistical Computations, Fizmatgiz, 1961.

THE WAVE FORM OF A LONGITUDINAL
REFLECTED WAVE FROM A THIN HIGH-SPEED LAYER

V. V. Kun

The wave forms for longitudinal waves reflected from a thin but uniform layer have been studied theoretically [1-5], on models [6-7], and in actual media [2, 4]. Recently, theoretical studies have appeared which develop the wave forms for waves reflected from a periodic sequence of thin layers [8-9].

The possibility of using the frequency spectrum of a wave reflected from a thin layer in the direct interpretation of seismic data has been examined in [1, 3, and 10]. In all cases, the spectrum has been studied only for normal incidence. The most thoroughly studied case is that for waves reflected from a thin uniform layer [1, 2, 4, 5]. A few experimental data on the forms of reflected waves have also been obtained. Comparison of such data with the corresponding theory has only just begun. In reference [1], recorded wave forms and amplitude spectrums were analyzed for cases of a very thin, high-speed layer and a thin, low-speed layer. However, satisfactory information about the velocity section was not obtained in these studies (there were no acoustic logs).

The present paper gives the results of an experimental and theoretical study of recorded wave forms and frequency spectrums for high-amplitude reflected waves recorded in the northern part of the Krasnodar Belt. By comparing the experimentally observed wave forms with those for waves computed by theory for various combinations of interval velocities taken from acoustic well logs, we have determined that these reflections correspond to an inhomogeneous, high-speed thin layer; we have also elucidated the nature of the effect of bed boundaries above and below this layer on the wave-form characteristics of the corresponding waves. This work is a first step in the overall project of constructing representations for actual media, as pointed out in the Introduction to this collection of papers. The differences between this study and earlier studies devoted to reflections from thin layers are as follows.

1. The recorded data were obtained at well sites where the velocity profile is well known from acoustic logs.

2. The high-speed layer is characterized by varying degrees of internal inhomogeneity, so that it is possible to study the effect of this inhomogeneity on the recorded wave forms.

3. The study of wave characteristics was carried out over a much broader frequency range, inasmuch as the spectrum of the incident waves (direct) was broader than the spectrum for the direct waves obtained with the field conditions described in [2, 3]. Moreover, the reflected waves were recorded not only at the surface of the Earth, but also at the base of the low-speed layer, which allowed the spectrum of the recorded waves to be extended to higher frequencies.

TABLE 1

Event	Well	Total l, m	$\overline{V}$, m/sec	f_{max}, cps	$\frac{l}{\lambda_3}$	$\frac{V_1}{V_2}$	$\frac{V_2}{V_3}$
$P_{0.3}$	52	10.5	2600	65	0.26	0.74	0.70
	(39)	11.0	2280	67	0.32	0.83	0.85
	36	$\{$ 11.5	2330	65	0.32		
		$\{$ 6.0	2730		0.14	0.70	0.68
$P_{1.3}$	52	24.5	5090	50	0.24	0.49	0.59
	39	$\{$ 33.5	4466	40	0.33	0.53	0.66
		$\{$ 36.0	4140	45	0.40	0.58	0.71
	45	$\{$ 60	3800	40—45	0.67—0.70	0.72	0.92
		$\{$ 13.5	4890	60	0.16	0.56	0.84

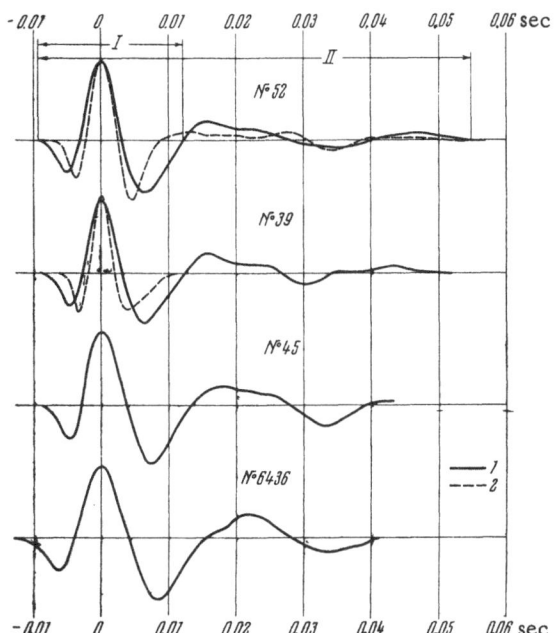

Fig. 1. Stacked records for the direct waves recorded at wells 52, 39, 45, and 6436, at the Earth's surface (1) and at the base of the low-speed layer (2). I and II are the time windows over which spectrums were computed.

4. In computing synthetic seismograms for comparison with experimental data, we used the recorded form of the direct wave as the incident pulse.

Method of Studying Waves

Field Survey Methods. Work on the study of strong reflections was carried out in the Staro-Minsk and Leningrad regions of the Krasnodar Belt, at four well sites where acoustic well logs were available (Wells 52, 39, 36, and 45). Surveys were carried out at a distance of 100-200 m on the surface from the well head and at the base of the low-speed layer using the so-called "seismic sounding method." In the seismic sounding method, the geophones are placed along two short spreads lying in a single vertical plane, with one spread at the surface, and the other at the base of the low-speed layer, in shallow drilled holes. Records are made from these spreads using one or two shot points located at a 30-50 m distance along the profile. Shots were fired at 1-5 m intervals at depths between 50 and 100 m, and by so doing, ghosts reflected from the bottom of the low-speed layer were not recorded. The shot size was held constant, being 0.3 kg (in 1963) or 0.4 kg (in 1964). As was pointed out in the preceding paper, such methods permit statistical stacking of the experimental data to improve the precision of the recorded wave forms and spectrums by a factor of 3 to 4 in comparison with single records. In recording, broadband filter settings were used, with no mixing, time shifting or automatic gain control being used.

With all the seismic soundings, the direct wave (P_1) was recorded with uphole phones at the shot point both at the surface and at the base of the low-speed layer, as suggested by V. V. Kuznetsov [16].

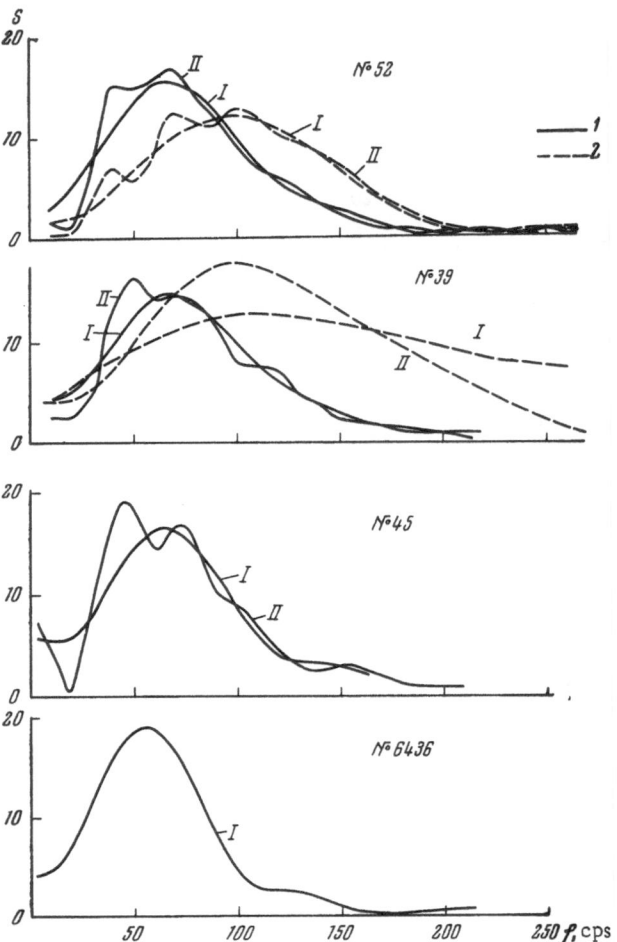

Fig. 2. Spectrums for direct waves, determined over two time intervals (I, II) for the stacked P_1 records shown in Fig. 1. 1) Spectrums determined from records at the surface; 2) spectrums determined from records at the base of the low-speed layer.

Theoretical Methods. The synthetic seismograms were calculated on a computer using a program written by S. S. Pariiskii and N. G. Mikhailov, based on plane-wave theory and following the approach of Baranov and Kunetz [9]. In this program, the reflected waves are evaluated at the Earth's surface. The stacked form of the direct wave P_1, also recorded at the surface, was used as the incident wave form. This allowed consideration of the effect of the travel path between the shot point and the surface on the reflected wave, inasmuch as this effect would be the same for the direct and reflected waves, for normal incidence. However, the effect of effective attenuation on the reflected wave is not considered in this method.*

The amplitude spectrum was obtained as the Fourier transform of the pulse seismogram.

---

*We will call the total action of such factors as spreading, scattering, and loss the effective attenuation.

TABLE 2

Well No.	h LVL, m	T_{02}, m/sec	T_{13}, m/sec	T_{24}, m/sec	$\dfrac{A_{12}}{A_{23}}$	$\dfrac{A_{34}}{A_{23}}$	$\dfrac{T_{02}}{T_{13}}$	$\dfrac{T_{24}}{T_{13}}$	f_{max}, cps	Δf, cps		Δf_{max}, %	
										0.7	0.5	0.7	0.5
52	6—7	8.5	11.6	14.5	0.81	0.58	0.73	1.26	68	57	79	84	116
39	11	8.8	11.0	15.3	0.85	0.57	0.80	1.38	70	56	84	80	120
45	6—7	8.8	11.8	18.3	0.87	0.57	0.75	1.55	65	52	72	80	111
36	15.5	11.6	14.7	21.7	0.79	0.64	0.82	1.48	56	45	62	86	110

Brief Review of the Characteristics of Inhomogeneous Layers

The acoustic logs of the intervals included in the high-speed inhomogeneous layers ($d_{0.3}$ and $d_{1.3}$) have been described earlier for the northern part of the Krasnodar Belt; a velocity section for these intervals, obtained from acoustic logs, is given in Fig. 2, p. 124. The depths to these layers are consistent over the region, being 280-300 m for the $d_{0.3}$ layer and 1300 m for the $d_{1.3}$ layer. Table 1 contains information about the characteristics of the $d_{0.3}$ and $d_{1.3}$ layers at wells 52, 39, 36, and 45, where the seismic sounding surveys were carried out. The thickness l of a layer, and the average velocity $\overline{V}$ in it were determined from the acoustic logs; the dominant frequency f_{max} was determined from the spectrum, and the wavelength was taken as $\lambda_2 = \overline{V}/f_{max}$. The $d_{0.3}$ and $d_{1.3}$ layers are most homogeneous at well 52, several sharp boundaries may be recognized within the layers at well 39, and at well 45, the $d_{1.3}$ layer has its greatest thickness and consists of several individual beds (Fig. 2, p. 124).

The lower values for l given in Table 1 for wells 36, 39, and 45 correspond to the beds with the highest wave speeds, while the higher values are for these beds combined with beds lying above (wells 36 and 39) or below (well 45) in the section.

Wave Form of the Direct Wave

Examples of stacked direct waves recorded at the surface and at the base of the low-speed layer are shown in Fig. 1. A wave P_1^{kp} recorded at the tail end of the record is a multiple reflection propagating through the low-speed layer, being reflected between the surface and the base of the low-speed zone.

Spectrums computed for short intervals of the stacked event P_1 (the interval I in Fig. 1) obtained at the Earth's surface are characterized by simple form and a comparatively large absolute and relative width (curves 1 on Fig. 2, and Table 2). Spectrums determined for the entire recording interval for the P_1 event (curves 2 in Fig. 2) repeat the general form of the spectrums determined over the interval I, but subsidiary maximums and minimums, reflecting the effect of the P_1^{kp} wave (the effect of the low-speed layer), are present.

The spectrums determined from records of the P_1 event recorded at the base of the low-speed layer are much broader than the spectrums for records at the surface (the Δf is larger by 34%) and the maximum is shifted towards higher frequency by 30 cps.

An attempt was made to find a relationship between recorded wave forms or spectrums for the P wave and the recording locations. A comparison of the stacked wave forms and spectrums for the direct waves obtained at various places on the Earth's surface (at wells 39, 52, 45, and 36) indicated that the P_1 wave forms and spectrums were consistent with respect to area. This may be seen in Figs. 1 and 2 and in Table 2, where the characteristics for the P_1 wave forms and spectrums are listed.

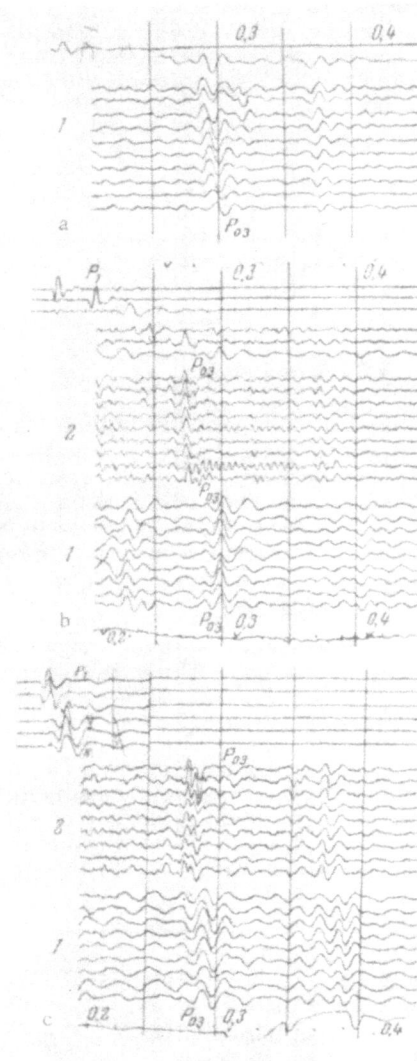

The consistency of the form of the P_1 pulse with respect to area is evidence, apparently, that the spectrum of the shot does not change much from area to area, and that the effect of the weathered layer on the transmission of P_1 waves through it is nearly the same in all areas studied, despite variations in the thickness of the low-speed layer.

Recorded Form of the Reflected Waves

Reflections on the seismic records were divided into two groups on the basis of amplitude (Figs. 3, 4). The dominant waves in each group are waves recorded at times of 0.3 and 1.3 sec, corresponding to the $d_{0.3}$ and $d_{1.3}$ beds. We will designate these as the $P_{0.3}$ and $P_{1.3}$ events.

The $P_{0.3}$ event corresponds to beds with essentially the same values for l and l/λ_2 ($l/\lambda_2 \approx 0.25$, Table 1) at the various well sites. The recorded form at the surface stays generally the same over the different areas (Fig. 5a). The maximum variation in T does not exceed 0.002 sec; the ratios of amplitudes of the main phases lie within a narrow range, from 0.6-0.8 (see Table 3). However, for some of the parameters, these variations are more than double the standard deviation contributed by measurement errors (see for example the values for 2σ for T_{24} and A_{34}/A_{23} at wells 39 and 52, columns 4 and 8 of Table 3*).

Thus, it may be stated with some degree of certainty ($p = 0.95$) that these variations do not result from experimental error.

The wave form $P_{0.3}$ obtained at the base of the weathered layer has a markedly higher-frequency character, and so, provides better resolution than the same event recorded at the surface (Figs. 3, 5b). Despite the effects of interference, its character changes only slightly between seismograms (Fig. 3b). The difference in recording time for the two most intense phases of the $P_{0.3}$ event (maximum No. 2 and minimum No. 3), even when confused by inter-

Fig. 3. Seismic records of the $P_{0.3}$ event obtained with a shot depth of 80 m. a) Well 6436; b) well 39; c) well 52; 1) at the surface; 2) at the base of the low-speed zone.

ference (Fig. 5b) averages 0.0080 sec in both areas. The value computed theoretically for Δt between waves from the top and the base of the $d_{0.3}$ layer is 0.0081 sec at well 52 and 0.0096 sec at well 39 (computed with the values for l and $\bar{V}$ listed in Table 1).

The coincidence between the experimental and theoretical values for Δt obtained at well 52 and the slight difference found at well 39, considering the simple, compact form of the incident wave, allows us to assume that the primary maximum of the P_1 wave recorded at the base of the low-speed layer (extremal 2, Fig. 5b) corresponds to the return of maximum energy from the top of the high-speed layer, while the succeeding minimum which is complicated by interference (extremal 3, Fig. 5b) is related to waves returned from the base of the layer.

*Values for $2\bar{\sigma}$ were determined approximately, from results in the preceding paper.

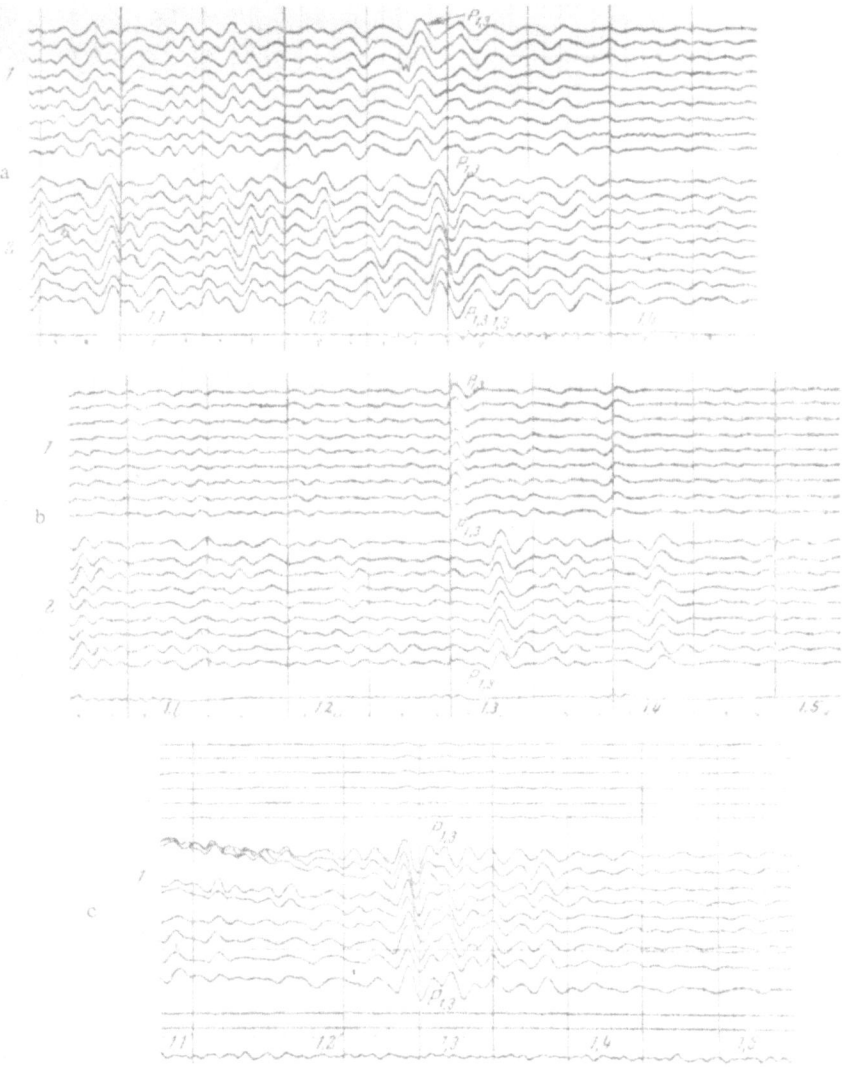

Fig. 4. Seismic records of the $P_{1.3}$ event obtained with a shot depth of 90 m. a) Well 52; b) well 39; c) well 45; 1) at the base of the low-speed layer; 2) at the surface.

The $P_{1.3}$ event is present with interference for a time intervals of 0.110-0.120 sec, and cannot be separated clearly from the waves immediately before and after it (Fig. 4). The recorded wave form does not vary from well to well (Fig. 6 and Table 3). At well 52, where the thickness of the layer is least and $l/\lambda_2 \approx 0.24$ (see Fig. 2, p. 124), the recorded $P_{1.3}$ wave form is very simple and symmetrical (Fig. 6). The periods T_{13}, T_{24}, and T_{35} are practically equal; the amplitude ratios A_{12}/A_{23} and A_{34}/A_{32} vary slightly. With increasing bed thickness (well 39, where $l/\lambda_2 \approx 0.40$), the periods T_{24} and T_{35} increase markedly in comparison with T_{13}. In thicker and more inhomogeneous parts of layer $d_{1.3}$ (at well 45, where $l/\lambda \approx 0.70$), the period T_{13} of $P_{1.3}$ is very short, and the second maximum (extremal 4) shows marked interference. With a further increase in l/λ_2 and the degree of inhomogeneity for the $d_{1.3}$ layer (that is, in going from well 52 to well 45), the relative intensity of the late phases of the $P_{1.3}$ event increases in comparison with the intensity of the first phase. Thus, at well 52 the ratio of amplitudes A_{45}/A_{23} is 0.20, at well 39 it is 0.42, at well 45 it is 0.62 (see Table 3).

TABLE 3

Event	Well	T_{02}, msec	T_{13}, msec	T_{24}, msec	T_{35}, msec	$\dfrac{A_{01}}{A_{23}}$	$\dfrac{A_{12}}{A_{23}}$	$\dfrac{A_{34}}{A_{23}}$	$\dfrac{A_{45}}{A_{23}}$	$\dfrac{T_{02}}{T_{13}}$	$\dfrac{T_{24}}{T_{13}}$	$\dfrac{T_{35}}{T_{13}}$
$P_{0.3}$	52	13.8	14.3	15.2	16.2	0.39	0.75	0.82	0.36	0.96	1.06	1.13
	39	11.2	12.4	17.2	19.5	0.42	0.87	0.63	0.33	0.90	1.39	1.57
	6436	12.3	14.5	15.7	15.6	—	0.77	0.65	0.25	0.85	1.08	1.08
$2\bar{\sigma}\,(P_{0.3})$	—		0.8	0.4	0.8	—	—	0.06	0.08	—	—	—	—
$P_{1.3}$	52	25.0	19.0	20.0	19.0	0.43	0.70	0.79	0.20	1.32	1.05	1.00
	39	25.0	19.0	27.0	28.5	0.44	0.82	0.63	0.42	1.32	1.42	1.50
	45	21.0	16.0	27.5	28.0	0.46	0.75	0.73	0.62	1.31	1.72	1.75
$2\bar{\sigma}\,(P_{1.3})$	—		0.6	—	1.2	—	—	0.06	0.06	—	—	—	—

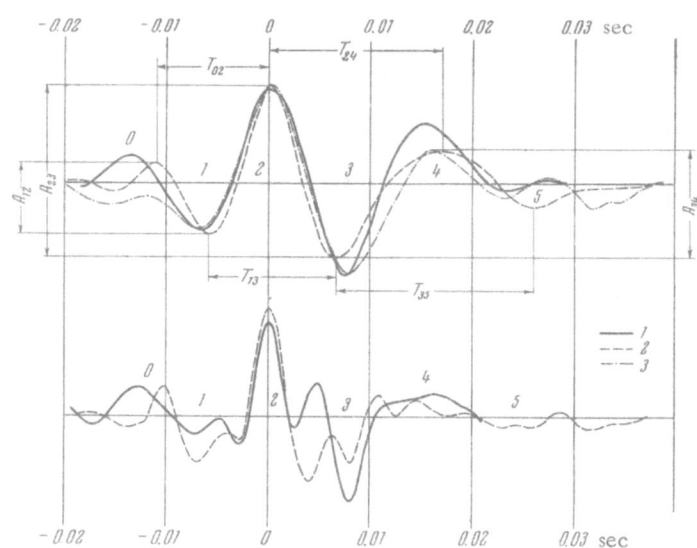

Fig. 5. Stacked records for the $P_{0.3}$ event, obtained from seismic sounding surveys. a) Surface recordings; b) recordings at base of low-speed zone; 1) well 52; 2) well 39; 3) well 6436. The numerals indicate the sequence of primary extremals.

The $P_{1.3}$ wave forms recorded at the base of the low-speed zone are practically the same as those recorded at the surface. This is explained by the fact that the $P_{1.3}$ wave has already lost its high-frequency components when it arrives at the base of the low-speed layer.

Spectrums for the $P_{0.3}$ and $P_{1.3}$ Waves

Wave spectrums were determined for several time windows for the stacked $P_{0.3}$ and $P_{1.3}$ events, inasmuch as the exact beginnings and endings of the waves were not known. Moreover, use of various spectral windows made it possible to evaluate the effect of boundaries in the sections above and below the $d_{0.3}$ and $d_{1.3}$ layers on the properties of the spectrums.

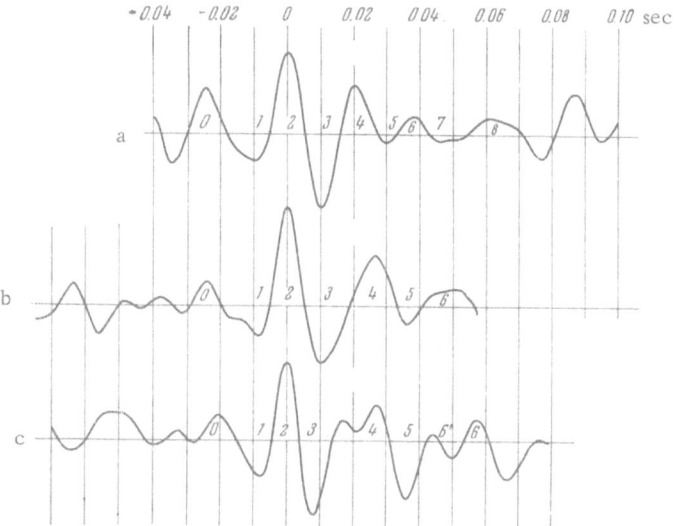

Fig. 6. Stacked wave form for the $P_{1.3}$ event, recorded
at the surface. a) Well no. 52; b) well no. 39; c) well
no. 45; d) the numerals indicate the sequence of phases
or extremals.

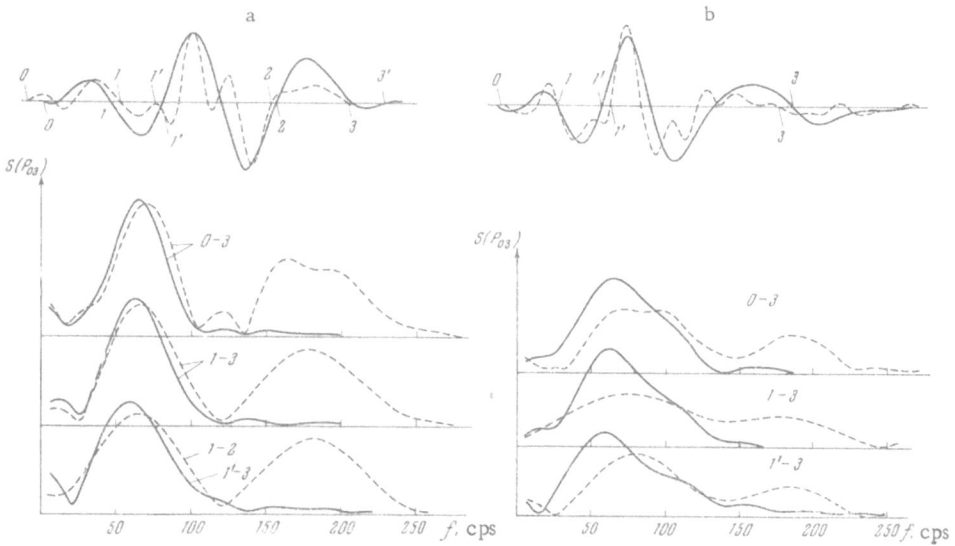

Fig. 7. Examples of spectrums for the $P_{0.3}$ wave, determined for various time windows for stacked records. a) Well 52; b) well 39. To the
left, surface measurements; to the right, measurements at the base of
the low-speed layer.

Spectrum of the $P_{0.3}$ Wave. The number of primary maximums and minimums
in the spectrum does not change as the time window for the original record is changed. The
frequencies at which the extremals are located shift slightly (Fig. 7). Consequently, the primary contribution to the energy of a wave is from the part of the section which forms the main
extremals on the record (the record interval 1-3, Fig. 5).

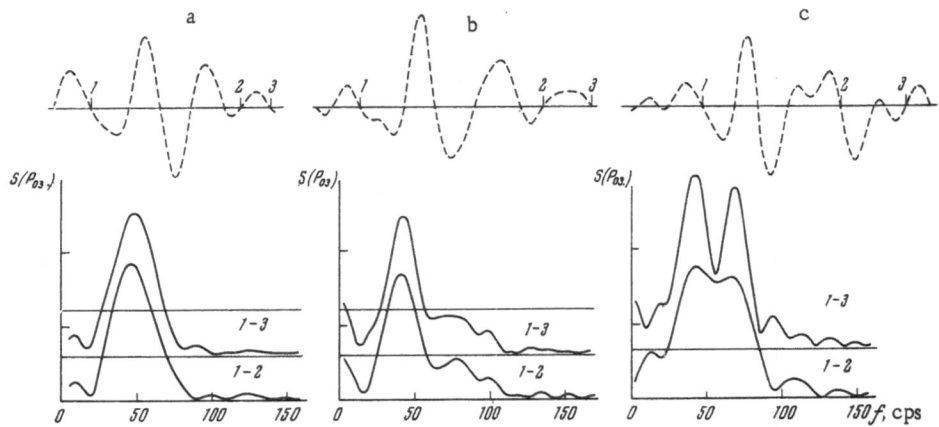

Fig. 8. Spectrums determined over two time windows from stacked records of the $P_{1.3}$ event, recorded at the Earth's surface. a) Well 52; b) well 39; c) well 45.

The spectrums for the $P_{0.3}$ waves, computed from stacked records made at the base of the low-speed layer at wells 52 and 39, are simple in form; there are two maximums with a minimum between them, which we will call the principal minimum.* At well 52, where the layer is most homogeneous, the amplitudes of both maximums in the spectrums are nearly the same, and their forms are symmetrical and simple. The properties of the spectrums differ only slightly from the properties of the spectrums for a homogeneous layer [1].† The difference in values for $2f_{max}-f_{min}$ and $\Delta f_{max\,1,2}-f_{min}$ amounts to 5-10 cps for values of $2f_{max\,1}$, $f_{min\,1}$, and $f_{max\,1.2}$ of the order of 120-140 cps.

For the spectrums at well 39, where the layer is less homogeneous, there is a diagnostically greater difference between the primary characteristics: $2f_{max\,1}<f_{min\,1}$ by nearly 15 cps, $f_{max\,1,2}<f_{min}$ by 25-30 cps, and the primary minimum is shifted towards higher frequencies (Fig. 7b).

The primary maximums and minimums for the $P_{0.3}$ wave spectrums determined from records made at the Earth's surface occur at practically the same frequencies as those for the spectrums determined from records made at the base of the low-speed layer (the difference amounts to 3-5 cps). The surface spectrums do differ in that the later maximum (at frequencies of 180-200 cps) is absent, which is explained by the greater attenuation of high-frequency waves on transmission through the low-speed layer.

In cases where the $P_{0.3}$ wave spectrum is determined for a time window which covers the first weak interference maximum (the interval 0-3 in Fig. 7a), secondary extremals appear on the primary maximums and the primary minimum splits in two; that is, the model becomes more inhomogeneous. These changes are explained by the addition of a thin low-speed layer to the $d_{0.3}$ bed from the overlying rocks.

Spectrums of the $P_{1.3}$ Waves. The $P_{1.3}$ spectrums were computed only for the records made at the Earth's surface. The character of the spectrums changes somewhat as the time interval for which they are determined changes; the minimums in the spectrums become sharper and deeper for longer time windows.

_____

*Here, as well as later, we define the first deep minimum which may be recognized following a maximum as the principal minimum. The nature of the minimum recognizable at frequencies near 20-30 cps is not clear, and we will not mention it further.

†For a homogeneous layer, $2f_{max}=f_{min}=\Delta f_{max\,1,2}$.

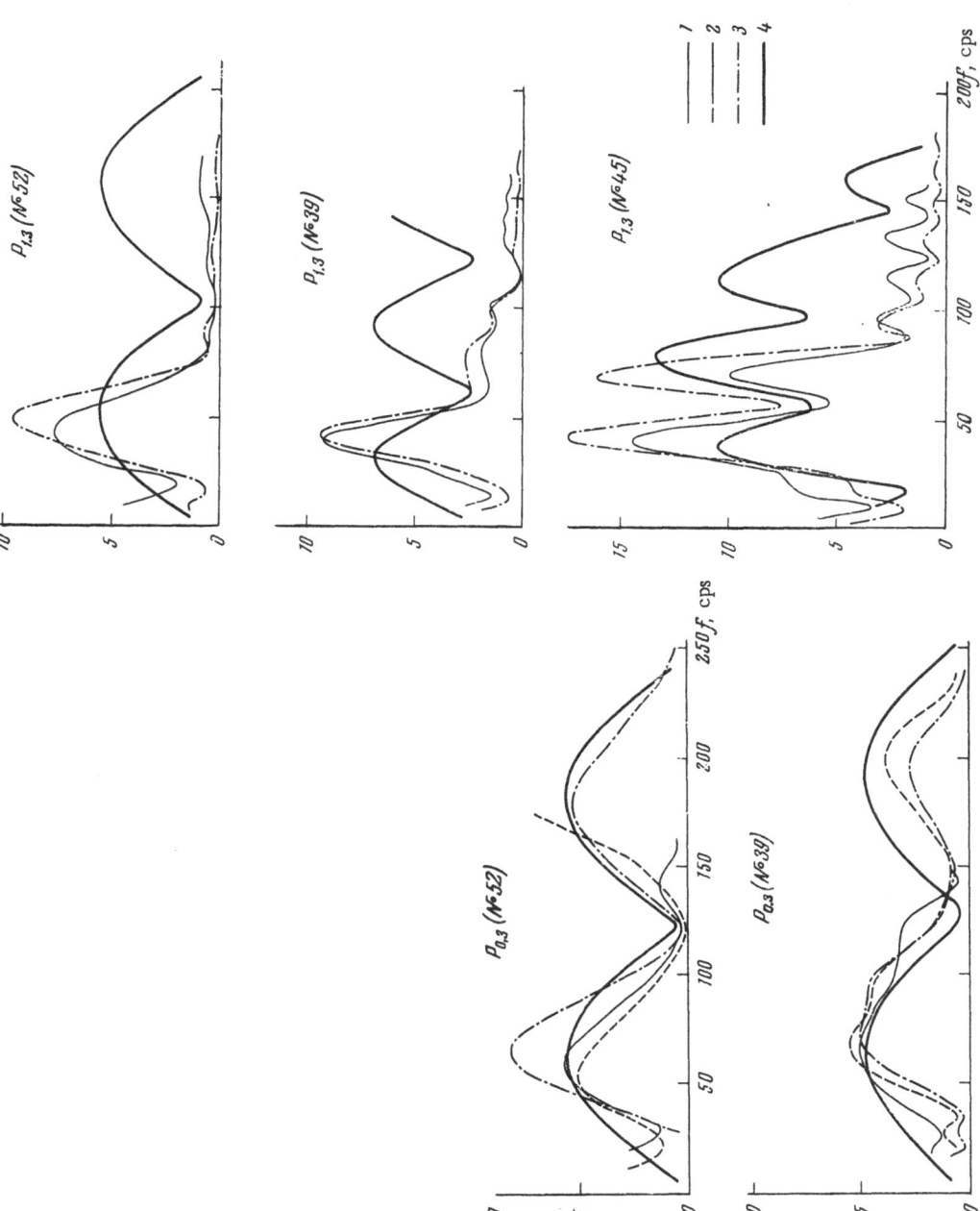

Fig. 9. Comparison of experimental spectrums (1) and spectral response functions (2, 3) with theoretical spectral response functions (4) computed for a uniform layer of thickness H. $P_{0.3}$, well 52, H = 10.5 m; $P_{0.3}$, well 39, H = 9 m; $P_{1.3}$, well 52, H = 24.5 m; $P_{1.3}$, well 39, H = 32 m; $P_{1.3}$, well 45, theoretical spectrums and response functions were determined for the interval 1-3 in Figs. 7 and 8.

TABLE 4

Source of l	No. of stacks	Record interval	f_{max}, cps	f_{min}, cps	$\Delta f_{max\,1,2}$, cps	l, m — f_{max}, cps	$\bar{f}_{min}$, cps	$\Delta f_{max\,1,2}$, cps	$\bar{l}$, m	$l_{acoustic}$, m	Notes
Well 52 ($\bar{V}$ = 2600)	19	0—3	140	104; 135	91	9.3	12.5; 9.6	13.8	10.5		$\bar{l}$ determined from spectrum
		1—3	134	120	112	9.9	10.8	11.6	10.8	10.5	
Base of low-speed layer	24	0—3	130	100; 137		10.0	11.8; 9.5		9.8		
Surface		1—3	126	122		10.3	10.6		10.5		
Well 39 (V =2280)	12	0—3	140	140	110	8.1	8.1	10.4	8.9		The same
		1—3	130	155	118	8.8	7.4	9.7	8.6		
Base of low-speed layer	9	0—3	150	150	125	7.6	7.6	9.1	8.1	11.0	
		1—3	140	155	130	8.1	7.4	8.8	8.1		
Surface	9	0—3	144	135		7.9	8.5		8.2		
		1—3	134	142		8.5	8.0		8.2		
Well 52	19	0—3	130	105; 133	—	10.0	12.4; 9.8	—	9.9		l determined from spectral response characteristic
Base of low-speed layer		1—3	108	120		12.9	10.8		11.5	10.5	
Surface	24	0—3	120	110; 140		10.8	11.8; 9.3		10.0		
		1—3	126	122		10.3	10.6		10.5		
Well 39	12	0—3	134	148	133	8.5	7.7	8.6	8.3		The same
Base of low-speed layer										11.0	
Surface	9	1—3	130	143		8.8	8.0		8.4		

With increasing thickness and inhomogeneity of the layer (in going from well 52 to wells 39 and 45),* the spectrums become more complicated; the number of maximums, as well as the level of high-frequency components, increases (Fig. 8).

Spectral Response Functions for the $d_{0.3}$ and $d_{1.3}$ Layers. The properties of the amplitude spectrums are determined not only by the spectral reflection function for a bed but also by the effects of shot hole conditions and geophone plant conditions as well as by the effective attenuation in the medium lying above the reflecting bed. The spectral response characteristic is controlled only by the spectral reflection function. However, the spectral response characteristic of the $d_{0.3}$ and $d_{1.3}$ layers cannot be determined accurately from the data described here because the necessary properties of the overlying medium are not known (attenuation and scattering). The ratios of the wave spectrums for the $P_{0.3}$ and $P_{1.3}$ events to the spectrum of the direct wave P_1 may be taken as the spectral response characteristic, as a first approximation; the effects of the spectrum of the shot, the response of the geophones, and attenuation in the medium above the shot point are excluded from these ratios; the effective attenuation in the medium between the shotpoint and the $d_{0.3}$ and $d_{1.3}$ layers is not excluded.

Curves for the ratios of the spectrums for the $P_{0.3}$ and $P_{1.3}$ waves to the P_1 spectrum are shown in Fig. 9. The spectrums for the $P_{0.3}$ and $P_{1.3}$ waves used in calculating the spectral response characteristics are shown on these same graphs for comparison. For all of the layers except the $d_{0.3}$ bed at the well 52, the experimental spectrums and the spectral response characteristics are very similar in form. The only differences are a lowering of the frequency of the first maximum in the spectral response characteristic in comparison with the spectrum, and a slight increase in the level of the high-frequency components for the spectral response characteristics. The slight differences between the spectral response characteristics and the spectrums may be explained primarily by the fact that the maximums in the spectrums for the direct and reflected waves are close, as well as by the simple form of the spectrum for the direct waves to the right of its maximum.†

---

* According to acoustic log data, the $d_{1.3}$ layer is more inhomogeneous at wells 39 and 45 than at well 52 (see Fig. 2, p. 124).

† A marked difference between the spectrum and the spectral response for f > 150 cps is found at well 52 for the $d_{0.3}$ bed. The reason for this is not known.

On the basis of this consideration of the properties of spectrums and response characteristics, we may conclude that a comparison of one of the parameters such as $f_{max\,1,\,2}$, f_{min},
or $2f_{max}$ allows a quantitative evaluation of the degree of inhomogeneity of a layer.

On the basis of this analysis, we may make some generalizations about the effects of inhomogeneity and thickness of a bed on the characteristics of a wave.

1. For layers with similar values for l/λ_2 ($d_{0.3}$; H = 280-300 m; $P_{0.3}$ wave), the recorded
wave forms at various observation points are similar with respect to the number and form of
extremals. Differences in the internal structure of a layer are reflected in slight differences
in several of the record parameters (T_{34}, A_{34}/A_{23}, and others) which exceed the errors in their
measurements.

The spectrums for the waves corresponding to such beds are also similar with respect
to the number of primary extremals and their distribution along the frequency axis. The inhomogeneity of a layer is reflected primarily in small irregularities in the distribution of the
primary maximums and minimums along the frequency axis.

2. For layers with significant differences in l/λ_2 ($d_{1.3}$, H $\approx$ 1300 m, $P_{1.3}$), a significant
change in recorded wave form and the spectrum is noted with a change in l/λ_2 or the internal
structure of the layer. With an increase in l/λ_2 or the degree of inhomogeneity of the layer, the
length of the recorded wave form and the intensity of the later phases are increased; the number
and size of extremals in the spectrum are also changed.

Evaluation of the Effective Thickness of the $d_{0.3}$ Layer. It was indicated earlier that the positions of the primary maximums and minimums of the spectrum
or spectral response characteristic for the $d_{0.3}$ layer are determined primarily by the positions of the top and bottom of the layer (the bed thickness); the internal boundaries of an inhomogeneous layer have less effect.

We have determined an effective thickness l from the properties of the spectrum and the
spectral response characteristic for the $d_{0.3}$ layer using the expression which is valid for a
uniform layer:

$$l = \frac{\overline{V}_2}{4f_{max\,1}} = \frac{\overline{V}_2}{2f_{min\,1}} = \frac{\overline{V}_2}{\Delta f_{max\,1,2}}$$

and compared it with the bed thickness found from acoustic logging data. The values for $\overline{V}_2$
were equal to $\overline{V}$ in Table 1. The difference between values for l found from the spectrums and
from acoustic logs may be considered to be roughly a quantitative measure of the degree of inhomogeneity of the $d_{0.3}$ layer (characterizing the reasonableness of representing a bed as being
uniform).

The determinations of l are listed in Table 4, where the index l indicates the arithmetic
average for the three determinations, from $f_{max\,1}$, $f_{min\,1}$, and $\Delta f_{max\,1,2}$.

The value for $\overline{l}$ determined from the spectral response characteristic (or spectrums)
practically coincides with the acoustic log data for the $d_{0.3}$ layer at well 52, but is smaller by
3 m at well 39. We may therefore conclude that the $d_{0.3}$ layer can be assumed to be uniform at
well 52, insofar as thick determinations are concerned, but that such an assumption would be
in error at well 39.

Synthetic Seismograms and Their Comparison with the Actual Records

Most of the work involving the use of synthetic seismograms has been straightforward.
Synthetic seismograms are used primarily for separating multiple reflections, so that the multiples may be excluded in interpretation.

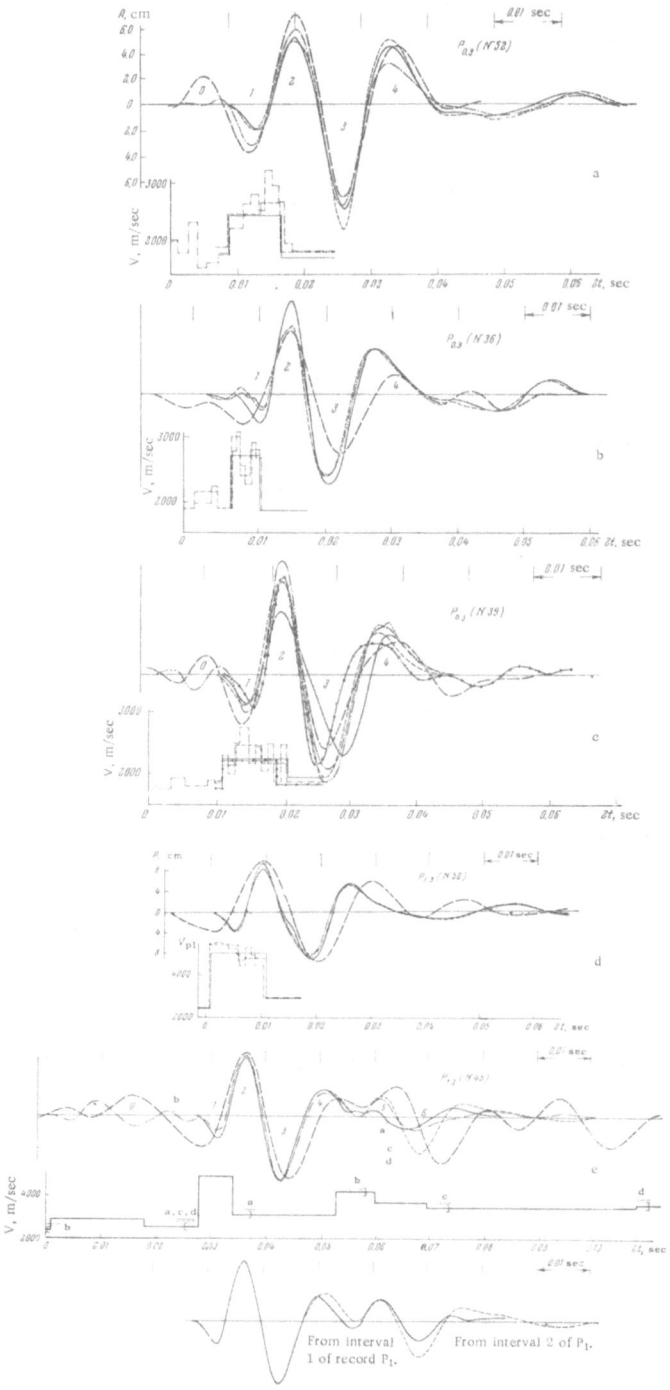

Fig. 10. Examples of synthetic seismograms computed for
various models of types I (a-d) and II (e). The velocity profiles for each model are shown under each group of records.
The designations for the seismograms and the sections are
the same; the numeral indicates the sequence of the extremals, and the simple dashed line indicates an experimental
seismogram.

In the present work, synthetic seismograms were computed for several models of different intervals shown on the acoustic logs, including the inhomogeneous layers $d_{0.3}$ and $d_{1.3}$. Comparison of these seismograms with one another permits theoretically an evaluation of the effect of various boundaries for the inhomogeneous layers on the recorded wave forms, and comparison of the synthetics with the stacked experimental seismograms permits the choice of the most likely representation of the actual medium.

We will compare synthetic seismograms among themselves and with experimental records for two types of models: 1) the $d_{0.3}$ and $d_{1.3}$ layers are represented in the form as actually shown by the sonic log, in averaged segments from the sonic log (with 2-5 layers) and as a uniform layer (see Fig. 10a-d); in going from the detailed log to one with 2-5 layers, only the stronge boundaries are kept, and the average velocities for the log segments represented in various ways are not changed; and 2) models were made from averaged acoustic log segments with a consecutive increase in the depth interval used for the computations (Fig. 10e).

The overlying and underlying beds were assumed to be homogeneous, as were the beds within the sequence (for models of types 1 and 2).

Synthetic Seismograms. The synthetic seismograms are shown in Fig. 10. The velocity profiles are shown below each group of synthetic seismograms, plotted to a time scale (the nomenclature is the same for the synthetic seismograms and the velocity sections).

The stacked records of the direct wave recorded at wells 52, 39, 6436, and 45 (see Fig. 1) were used as the incident wave forms in computing the synthetic seismograms. When the entire wave form for the P_1 wave is used (interval 2 in Fig. 1), the effect of the low-speed layer on the transmission of the reflected wave is considered completely, inasmuch as the complete wave includes the wave component P_1^{kp}.

A comparison of the seismograms computed for intervals 1 and 2 of the wave form P_1 indicates that the difference is small for the first three phases (extremals 1-3, Fig. 10e'). Consequently, one may use the shorter interval for the P_1 wave (interval 1) in computing seismograms for reflected waves of moderate length. The latter parts of the reflected wave forms computed using interval 1 and interval 2 from the P_1 wave differ significantly (Fig. 10e'). Therefore, if the latter phases are to be studied, the full wave form for the P_1 wave must be used (interval 2).

We may draw the following conclusions from a comparison of the seismograms computed with various models of type 1, for a single well.

1. For the case of an inhomogeneous layer in which the contrasts in propagation speeds at the interval boundaries are moderate (the reflection factors at the internal boundaries are less than the reflection factors at the top and bottom of the bed), the top and bottom boundaries exert the strongest influence on the recorded wave form. The acoustic log interval containing such a layer may therefore be represented as a uniform layer, with top and bottom corresponding to the boundaries of the beds with the highest propagation speeds.

This in quite obvious from Fig. 10a and 10d, where there is close agreement between the synthetic seismograms computed for all three variants of the model for an inhomogeneous layer with weak boundaries internally (the detailed sonic log model, the 2- and 3-layer models, and the uniform layer model).

This is obvious as well from Fig. 10b, where the synthetic seismogram computed for a less-uniform sonic log section with a thickness of 11.5 m and the synthetic for the same section split up into six layers are both very close to the seismogram corresponding to a uniform layer with a thickness of 6 m characterized by a much higher velocity.

TABLE 5

Wave	Well and type seismogram	T_{03}, m sec	T_{13}, m sec	T_{24}, m sec	T_{35}, m sec	$\frac{A_{12}}{A_{23}}$	$\frac{A_{34}}{A_{23}}$	$\frac{A_{45}}{A_{33}}$	$\frac{T_{02}}{T_{13}}$	$\frac{T_{21}}{T_{13}}$	$\frac{T_{35}}{T_{13}}$
$P_{0.3}$	52 (S)*		10.3—10.6	13.0—13.2	—	0.57; 0.59	0.78; 0.85	0.27; 0.23		1.26; 1.24	
	52 (F)	13.8	14.3	15.2		0.75	0.82	0.36		1.06	
	39 (S)	9.2	10.8	11.6—16.1		0.62—0.71	0.69—0.83		0.97	1.08—1.36	
	39 (F)	11.20	14.4 12.4	17.2	25.8—26.0	0.87	0.63	0.33	0.90	1.38	2.4—2.8
	33 (S)	7.5—7.8	9.2—10.6	12.3—13.0	15.5—32	0.54—0.66	0.74—0.86	0.34—0.45	0.71—0.85	1.16—1.41	1.07—2 2
	36 (F)	12.3	14.5	15.7		0.77	0.65	0.25	0.86	1.08	
$P_{1.3}$	52 (S)	12.0—26	15.0	17.5	19.0	0.68	0.70	0.28	0.80	1.17	1.10
	52 (F)	25.0	19.0	20.0	19.0	0.70	0.79	0.20	1.32	1.05	1.00
	45 (S)	14.1	11.6—12.0	24.5	23.8—28	0.61—0.66	0.66—0.69	0.26—0.47	1.17	2.05; 2.10	2.10—2.46
	45 (F)	21.0	16.0	27.5	28.0	0.75	0.73	0.62	1.31	1.72	1.75

*S = synthetic; F = field.

2. In cases in which the reflection factors at boundaries within a layer are the same as those at the top and bottom boundaries, the wave form for an inhomogeneous layer differs from the wave form for a uniform layer of the same thickness. The principal differences are in the lengths of the periods.

Thus, for example, there is little difference between the synthetic seismograms computed for the $P_{0.3}$ event at well 39 for various representations of the 11-m $d_{0.3}$ layer, but the seismogram computed for a homogeneous 11-m layer is quite different (Fig. 10c). Consequently, judging by wave form, a particular inhomogeneous layer may not be mistaken for a uniform layer of the same thickness.

Similar conclusions were drawn concerning the possibility of distinguishing an inhomogeneous layer from a uniform layer in reference [9], on the basis of theoretical frequency spectrums.

In cases in which the reflections are formed by several inhomogeneous layers (layer $d_{1.3}$ at well 45, as shown in Fig. 2, p. 124), the comparison of seismograms computed for models of type 2 allows a determination of the primary boundaries and thicknesses of the sequence of the layers which contribute to the formation of the reflection. Thus, for example, the beginning and main parts of all the synthetic seismograms for the $P_{1.3}$ wave at well 45 (Fig. 10e, extremals 1-4) are similar for all models beginning with the model for a uniform $d_{1.3}$ layer with a thickness of 14.7 m and the highest speed of propagation, and ending with the last models in the form of sequences of uniform beds with thicknesses of 78 and 145 m (models c and d, respectively). The amplitudes of the later phases of the $P_{1.3}$ wave (extremals 5 and 6 in Fig. 10e) differ for the different models; they increase as the total thickness of the sequence increases, but for the sequences with thickness of 78 and 145 m they are the same. Thus, the $P_{1.3}$ wave form at well 45 is determined by the makeup of the entire sequence of layers with a thickness of nearly 80 m but the principal effects on the most intense early phases are controlled by the first 15 m with the highest propagation speeds.

Comparison of Synthetic and Experimental Seismograms. Similar-

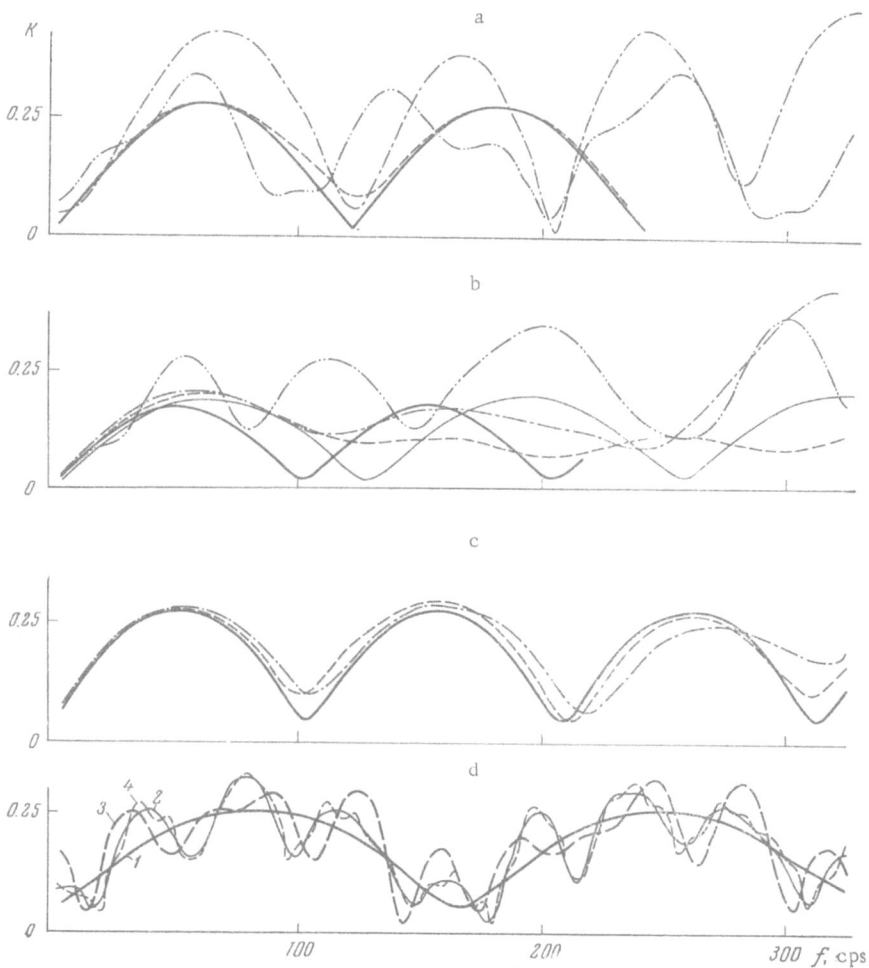

Fig. 11. Examples of theoretical amplitude response functions computed for various models of the $d_{0.3}$ and $d_{1.3}$ layers. The curve designations in a, b, and c are the same as those in Fig. 10; while d is explained in the text. a) $P_{0.3}$ at well 52; b) $P_{0.3}$ at well 39; c) $P_{1.3}$ at well 52; d) $P_{1.3}$ at well 45.

ities and differences between the experimental and synthetic seismograms have been established. Both these, and others, have similar numbers of primary phases, and a close correspondence between amplitudes (differences are of the order of 10-20%). This is quite apparent from Fig. 10 and the data in Table 5, where the characteristics of the experimental stacked seismograms are listed, along with the ranges for the parameters for the synthetic seismograms shown in Fig. 10.

The synthetic seismograms, as well as the experimental, are similar for the $P_{0.3}$ wave at all wells, but differ significantly for the $P_{1.3}$ wave.

The differences between the synthetic seismograms at the various wells are essentially the same as those between the field seismograms (see, for example, the transition from well 52 to well 45 in Table 5 — the periods T_{24} and T_{35} of the synthetic and field seismograms increase, T_{13} decreases, and so on).

These similarities make it possible to associate the various phases on a field seismogram with the section, using the synthetic seismograms as a guide. It is apparent from Fig. 10

that the onset of the $P_{0.3}$ and $P_{1.3}$ waves (minimum 1) corresponds to the top of the layer with high wave speed. This is quite apparent for all of the models with a uniform overburden above the reflector. The seismograms computed for models with nonuniformities in the overburden differ in that there is a weak interference maximum ahead of the first minimum phase (Fig. 10b, c, d).

There are systematic differences between the periods of the waves on the field and synthetic seismograms; the T_{02}, T_{13}, and T_{24} periods on the field seismograms are 1.1-1.5 times longer than on the synthetic seismograms. The most probable explanation of these differences may be the lack of consideration of the effects of attenuation in the medium between the shot point and the reflecting surface in the theoretical seismograms.

The existence of systematic differences between field and synthetic seismograms means that for the time being, models for the medium apparently cannot be selected on the basis of a comparison of field seismograms with synthetic seismograms.

Theoretical Amplitude Spectral Response and Comparison with Field Results

In reference [9], the amplitude spectral response function was analyzed for a periodic sequence of layers, and the variation of these functions with the number of layers in a periodic sequence and with the ratio of reflection factors between the layers was investigated; the possibility of distinguishing between a sequence and a uniform layer was investigated. The study was carried out for theoretical sections. In this reference, theoretical spectral response functions computed for the models of the $d_{0.3}$ and $d_{1.3}$ layers as were used for the synthetic seismograms in Fig. 10 were compared among themselves (Fig. 11) and with field seismograms (Fig. 9). This theoretically permits us to study the influence of various boundaries in the inhomogeneous $d_{0.3}$ and $d_{1.3}$ layers on the spectral response function and to apply inferences with the appropriate experimental results.

Theoretical Amplitude Spectral Response Function. It is apparent from Fig. 11a and b that the top and bottom of the sequence contribute the principal effect on the spectral response for inhomogeneous $d_{0.3}$ and $d_{1.3}$ layers which have only moderate internal contrasts in wave speed. The spectral response function computed for a model of an inhomogeneous layer which matches the acoustic log in detail has the same number of maximums and minimums as the response function for a uniform layer of the same thickness (Fig. 11a, c); the frequencies for the primary extremals for these two models either nearly coincide (Fig. 11c) or differ by a moderate amount (Fig. 11a). The difference between the functions consists of higher frequencies for the extremals, lesser depths for the minimums, and larger values for the reflection factor K for the response function computed for the acoustic log section.

The difference between response functions for models of inhomogeneous layers consisting of two- or three-bed sequences and for models of uniform layers are qualitatively the same as those described above, but to a lesser degree. Thus, for example, for the appropriate response functions shown in Fig. 11a and c, the difference is in the size of the reflection factor K about the minimum.

The response function computed for an inhomogeneous layer following the acoustic log in detail with strong internal boundaries is quite different from the response function for the corresponding uniform layer; the extremals are shifted by significant amounts, and the ratios of amplitudes and frequencies are changed (Fig. 11b).

The response functions computed for models with inhomogeneities in the overlying and underlying beds differ from the response functions for the acoustic log models or for uniform layers with a constant wave speed $\overline{V}$ in the surrounding rocks in that subsidiary minimums appear, usually with lesser depths than the primary minimums. Moreover, some shifts in the primary maximums and minimums take place. For example, the introduction of a uniform layer with low wave speed in the overlying beds into the model for the/$d_{0.3}$ layer at well 52 leads to a splitting of the primary minimum and all of the maximums (Fig. 11a).

Comparison of Theoretical and Experimental Response Functions

The experimental response functions for the $d_{0.3}$ layer at wells 52 and 39 and the $d_{1.3}$ layer at well 52, over a frequency range of 30-150 cps for the $P_{0.3}$ wave recorded at the base of the low-speed layer, and of 30-100 cps for the $P_{1.3}$ wave are similar to the response functions computed for the corresponding uniform layers. They are similar with respect to the number of primary maximums and minimums, their relative intensities, and their positions along the frequency axis (differences are of the order of 5-10 cps).

The experimental response function and spectrum obtained for the $d_{1.3}$ layer at well 45 over the frequency range 30-100 cps (Fig. 9e) have the same numbers of maximums and minimums and similar locations on the frequency axis as the theoretical characteristics computed for model c of Fig. 10e.

Inasmuch as it has been shown for the theoretical response functions that the positions of the primary maximums and minimums are determined quite accurately (well 52) or at least approximately (well 39) by the depths to the top and bottom of an inhomogeneous layer, this result may be extended to the corresponding experimental spectrums and response functions for the $d_{1.3}$ layer at well 52 (Fig. 9c) which allows us to draw the same conclusions for this inhomogeneous layer.

On the basis of a comparison of experimental and theoretical response functions at well 45, we may conclude that the basic aspects of the experimental response function are determined primarily by the effects of the tops and bottoms of the two inhomogeneous layers with high wave speed, as indicated on model c in Fig. 10e. The same results concerning the nature of the influence of internal boundaries in inhomogeneous layers are obtained theoretically and experimentally (see Fig. 7b and Fig. 11d, curve 2).

Layers and boundaries located above or below the $d_{0.3}$ or $d_{1.3}$ beds lead to the development of subsidiary extremals, complicating the primary maximums and minimums. Corresponding examples may be seen in Fig. 7, where the $P_{0.3}$ wave spectrum for the time interval 0-3 at well 52 is shown, and in Fig. 11a, where the response function for a model of a homogeneous $d_{0.3}$ layer at well 52 with a low-speed layer above is shown.

The width of the experimental response functions is less than that of the theoretical. This difference is greater for the $d_{1.3}$ layer than for the $d_{0.3}$ layer, and amounts to 50%. Inasmuch as the experimental and theoretical response functions are determined for essentially the same section interval, the basic reason for the differences in width cannot be a difference in thickness of the section being studied. It is more likely that a primary role is played here by the influence of effective attenuation, which was not included in determining the experimental response functions.

Conclusions

We have studied the recorded wave forms, the spectrums and the response functions for waves reflected at normal incidence from an inhomogeneous layer with high wave speed. We have presented synthetic seismograms and theoretical response functions, and we have compared these with experimental data.

The following results have been established:

1. The large amplitude reflections ($P_{0.3}$ and $P_{1.3}$) recorded at normal incidence in the northern parts of the Krasnodar Belt represent inhomogeneous thin layers with high wave speed; the values for l/λ_2 for these layers lie in the range 0.20-0.70.

2. The wave forms for the $P_{0.3}$ and $P_{1.3}$ waves depend on the structure of the inhomogeneous layer. For moderate wave-speed contrasts at boundaries within an inhomogeneous layer, the top and bottom boundaries of the high-speed beds play the primary role in determining the wave form. Evidence for this is an analysis of synthetic seismograms, which showed the similarity between the experimental response function and a theoretical response function computed for a homogeneous layer. Strong boundaries within a homogeneous layer lead to essential changes in the properties of the response function with respect to the response function for a homogeneous layer. Boundaries in the overlying or underlying rocks located close to an inhomogeneous layer lead to the appearance of weak subsidiary extremals on the response function for an inhomogeneous layer.

3. The experimental wave forms and spectrums for the $P_{0.3}$ wave, and especially for the $P_{1.3}$ wave, change in regions where there are changes in the thickness or internal structure of an inhomogeneous layer.

4. Comparison of experimental wave forms and response functions with the corresponding theoretical characteristics has shown that they are reasonably similar. However, there are significant differences between them: longer periods on the field seismograms and a lesser width to the response function. The cause of these systematic differences is apparently the effective attenuation in the medium between the shot point and the reflecting surface.

LITERATURE CITED

1.   I. S. Berzon, A. M. Epinat'eva, G. N. Pariiskaya, and S. P. Starodubrovskaya, Wave Forms of Seismic Waves in Actual Media, Izd. Akad. Nauk SSSR, Moscow, 1962.

2.   S. P. Starodubrovskaya, Physical basis for using wave forms of longitudinal reflected waves for tracing layers of varying thickness, Izv. Akad. Nauk SSSR, Ser. Geofiz., No. 12 (1964).

3.   S. P. Starodubrovskaya and G. N. Pariiskaya, Use of wave forms of reflected waves for identifying and tracing layers of varying thickness, Razved. Geofiz., No. 2 (1964).

4.   I. I. Gurvich, On reflections from thin layers in seismic exploration, Prikl. Geofiz., No. 9 (1952).

5.   I. I. Gurvich, Analysis of reflections from thin layers, Prikl. Geofiz., No. 15 (1956).

6.   L. A. Ivanova, Studies on models of the ratios of intensities of reflected and head waves, Tr. Inst. Fiz. Zemli, No. 30:197, Seismic Modelling.

7.   V. P. Telezhenko, Yu. B. Gorshenin, and L. I. Dorognitskaya, Wave forms of seismic records in the case of inclined layers from model data, Tr. Sibirsk. Nauchn.-Issled. Inst. Geol. Geofiz., i Mineral'n. Syr'ya, No. 27 (1962).

8.   V. S. Isaev, Results of the use of theory of grouping for studying periodic sequences of layers, Prikl. Geofiz., No. 49 (1967).

9.   N. G. Mikhailova, B. S. Pariiskii, and M. V. Saks, Response function of a sequence of layers, Izv. Akad. Nauk SSSR, Ser. Fiz. Zemli, No. 1 (1964).

10.   L. L. Khudzinskii, On determining some spectral characteristics of layered media, Izv. Akad. Nauk SSSR, Ser. Geofiz., No. 3 (1962).

11.   V. V. Kuznetsov, Determining the average attenuation factor and reflection factor from spectrums and amplitudes of direct and reflected waves, Tr. Inst. Fiz. Zemli, Akad. Nauk SSSR, No. 34:201 (1964).

12. Paul Wuenschel, Seismogram Synthesis Including Multiples and Transmission Coeffi-
     cients, Geophysics, 25(1):106-129 (1962).
13. N. A. Anstay, Attacking the Problems of the Synthetic Seismogram, Geophys. Prospecting,
     8(2):242-259 (1962).
14. P. Bois, J. Chauveau, and G. Grau, Sismogrammes synthetiques: possibilités, techniques
     de réalisation et limitations, Geophys. Prospecting, 8(2):260-298 (1962).
15. H. Durschner, Synthetic Seismograms from Continuous Velocity Logs, Geophys. Pros-
     pecting, 6(3):272-284 (1962).

HEAD WAVES FROM THIN LAYERS
(FROM FIELD DATA)

A.M. Epinat'eva and E.V. Karus

The study of seismic waves from thin layers is of considerable practical interest, inasmuch as thin layers with distributed elastic properties seems to be the normal mode of occurrence for actual media made up of sedimentary rocks.

The supposition that head waves may be formed on thin layers with high wave speed is the physical basis for the refraction method [1]. This supposition is based on differences in dilatational and boundary propagation speeds noted in the study of velocity sections from many areas [1, 2, and others]. Studies of the properties of head waves from thin layers have been carried out by modelling, by experiments in actual media, and by solution of the theoretical problem. These studies have become more sophisticated with time, and this has led to a significant change in physical concepts.

Quantitative attempts to determine the thickness of refracting layers which form head waves in actual media have been described in references [3, 4]. In one example (the fore-Baltic area, $H = 5$-15 m, $V_{P_1}/V_{P_2} = 0.4$) it was concluded that head waves could be recorded for very small ratios of bed thickness to wavelengths ($l/\lambda_2 \approx 0.06$). In another example (the fore-Carpathians, $H = 1300$ m, $V_{P_1}/V_{P_2} = 0.5$) head waves were recorded for $l/\lambda_2 = 0.15$.

The nature of head waves is quite different for the cases of thin and thick beds. In thin layers, an interference wave pattern is formed by the superposition of single-reflected and multiple-reflected waves in the layer and head waves [5]. The thinner the layer and the greater the wave length λ_2 in it, the more quickly are these waves formed.

Modelling with liquid/solid models [6] has shown that the wave forms of head waves change very markedly with variations in l/λ_2. In models with a solid sheet immersed in a liquid medium, head waves may be formed on very thin layers ($l/\lambda_2 < 0.1$) and yet have quite large amplitudes.

Head waves on very thin layers ($l/\lambda_2 \lesssim 0.18$) have practically the same amplitudes and attenuate slowly with distance, as do those on thick layers ($l/\lambda_2 \gtrsim 2$). For layers of intermediate thickness ($l/\lambda_2 = 0.7$-1.3) the waves are weak and attenuate rapidly with distance. On the basis of such work, it has been stated in one book [4] that under actual conditions, thin beds provide the most favorable circumstances for the application of the refraction method, while beds of intermediate thickness are less favorable.

Further investigations have shown that it is not possible to extend the results of modelling with liquid/solid models to actual media. Differences in the nature of the contact (slipping at a solid/liquid contact as opposed to elastic at a solid/solid contact) may explain the very significant differences in wave forms [7].

142

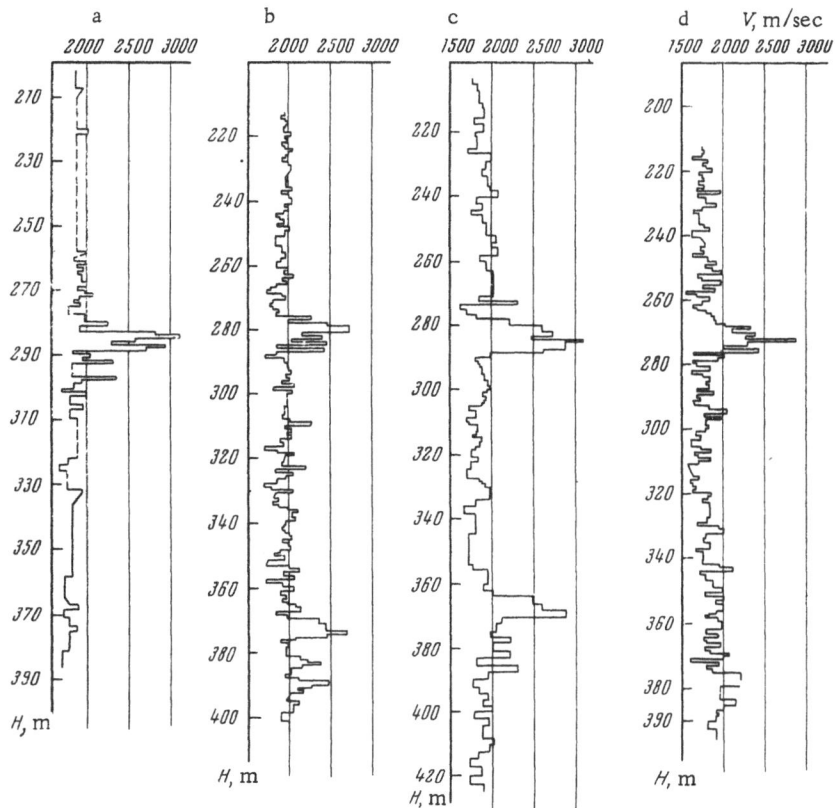

Fig. 1. Velocity section over the depth interval containing a thin layer with high wave speed (h = 280 m). a) Well 6436; b) well IV (near well 39); c) well 52; d) well 42.

Work on the study of head waves on two-dimensional solid models appeared in 1961-1963 (in 1961, the work of Laverne [8]; in 1962 the work of Laverne and Ingram [9]; and in 1963, the work of P. G. Gil'bershtein and I. I Gurvich [10]). Different types of modelling materials were used in the various studies, the contrasts in wave speed were different, the depths of layers varied, and so on, but a single type of result was obtained: with decreasing ratio l/λ_2, the amplitudes of head waves decreased and the rate of decay with distance increased.

In reference [8], the minimum value for l/λ_2 for which head waves were obtained was 0.083; in reference [9], it was 0.2. Observational results were obtained for still-lower ratios l/λ_2 in reference [10]. With $l/\lambda_2 = 0.06$, the head waves could not be identified, while for $l/\lambda_2 = 0.14$, they are recognizable only over a short interval.

Waves from thick and thin layers are compared in reference [11]. The ratios of amplitudes for waves of various types were essentially the same. High-amplitude PP and PS waves were recorded from a very thin layer ($l/\lambda_2 = 0.4$), being traceable to large distances ($x/H = 10$), but head waves were very weak. In the case of a thick layer ($l/\lambda_2 = 1.3$), the PP and PS waves could be identified only at short distances, but the head waves were strong; a number of dilatational refracted and multiply refracted waves with equal apparent velocities V* could be recognized.

Thus, modelling studies in solid media have shown that waves from thin layers have distinctive properties in respect to waves from thick layers; head waves are weak and decay more quickly with distance than in the case of thick layers, and the ratios of amplitudes of reflected and head waves also differ in the two cases.

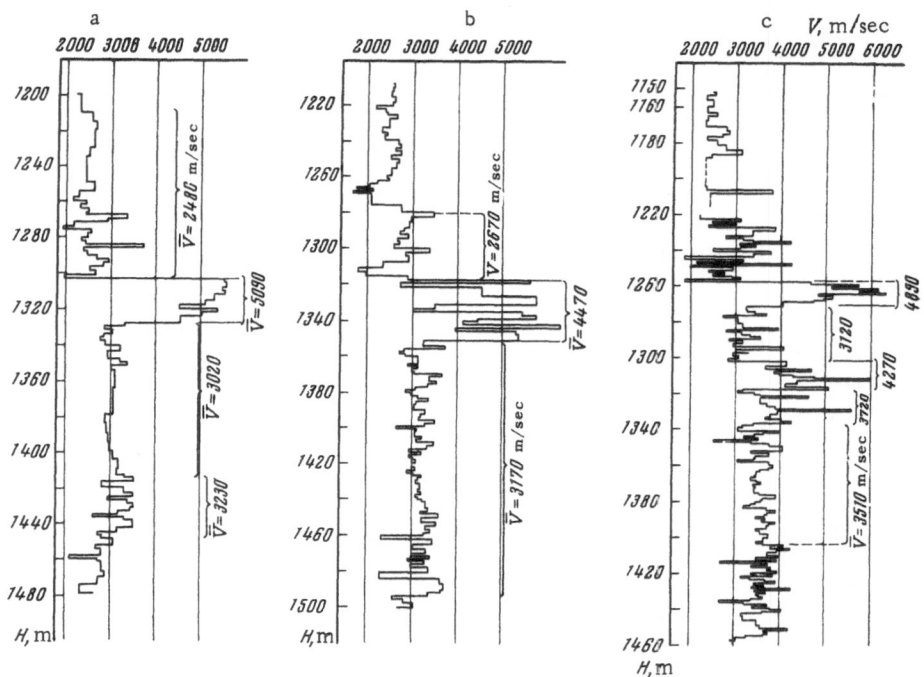

Fig. 2. Velocity section over the depth interval containing a sequence of layers with high wave speed (h ≈ 1300 m). a) Well 52; b) well 39; c) well 45.

In the model studies described in reference [12] for a layer of significant thickness, attempts were made to separate the head waves and waves reflected from the bottom of the layer, where the head wave was very weakly reflected. Starting with that distance at which the transit times for head and reflected waves are about the same, the reflected wave (being the more intense) determines the amplitude and arrival time for the interference wave propagating along the layer.

In theoretical work by Molotkov and Krauklis [13], it was concluded that it is impossible to have a low-frequency head wave on a thin layer (with $\lambda \gg l$). However, numerical results were not obtained, and so it is not possible to compare field results with the theory in a quantitative manner. Qualitatively the results are in agreement; the amplitudes of head waves from very thin layers are very small.

The results of modelling and of the solution of the theoretical problem for solid media are not in agreement with accepted ideas about the behavior of head waves from thin layers in actual media [1, 4]. However, the data on which these ideas are based is inadequate. Therefore, there is a real need for new data on head waves from thin layers in real media. The present paper explores this problem.* We present here the results of an experimental study of head waves from thin layers. The important feature of this study is the wealth of information available about the velocity section (from acoustic logging).

The work was performed by a field party from the Institute of Physics of the Earth of the Academy of Sciences of the USSR.

—————

*At the time this paper went to press, a paper by E. I. Gal'perin came to our attention: "On the intensities of head waves and surface-reflected waves based on data from vertical seismic profiling." (Izv. Akad. Nauk SSSR, Ser. Fizika Zemli, No. 10, 1966). Conclusions concerning the small amplitudes of head waves from thin layers were drawn on the basis of well surveys.

TABLE 1

No.	h, m	l, m	V_{P_2}	V_{P_1}	V_{P_3}	$\overline{V}_{P_1}$	$\overline{V}_{P_1}/\overline{V}_{S_1}$	V_{P_1}/V_{P_2}	$l\lambda_2$	H/λ_1
1	280	6—11	2200—2600	1800	1800	1600	3.5—4.0	0.69—0.82	0.12—0.32	8
2	1300	25	5090	2500	3000	2000	2.7	0.49	0.20—0.24	26
3	1260	13	4890	2700	3600	1900	—	0.51	0.13—0.16	26

Experimental Procedure

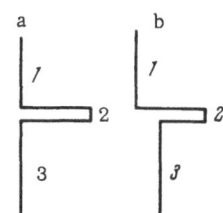

Fig. 3. Simplified models for velocity profiles.

Field work was carried out in the northern part of the Krasnodar Belt. To a depth of 1300 m, the section consists of sandstone and shale beds; these are underlain by a layer of limestone, and below this there are more sandstone and shale beds.

The velocity profile was obtained by acoustic logging in a number of wells, so that thin beds may be recognized and their wave speeds determined with an error of about 2%. Only the longitudinal propagation speeds were studied.

One of these velocity profiles is presented in Fig. 1 in the paper by Karus and Saks in this collection. Several layers with diagnostically high wave speeds can be identified in this section, and these were the target of the study.

The basic objectives of the study were: 1) a thin layer or sequence of thin layers at a depth of about 280 m; this zone is situated in a medium with very uniform wave speed, so that the wave speeds above and below the zone are practically constant; and 2) a sequence of thin layers at a depth of about 1260-1300 m; this sequence is more complex than the first zone, and the wave speeds above and below the sequence vary. The upper sequence is a group of sandstone and shale beds, and lies in a medium which also consists of sandstones and shales; the lower sequence consists of limestones and lies in a medium which consists of sandstone and shales.

The layer at 280 m depth can be seen in all wells in all wells in the various areas investigated, its thickness varies from 6-11 m, and the details of the velocity structure and the average wave speed over the whole of the layer sometimes differ markedly (Fig. 1, Table 1).

The layer with increased wave speed at a depth of about 1300 m also is present in the various areas. The layer thickness and wave speeds vary between areas (Fig. 2, Table 1). The wave speed in the underlying beds also varies. The wave speed is much higher over a considerable thickness of rock beneath the lower layer in one area (Fig. 2c) than in the other areas (Fig. 2a, b) (3500-3700 and 3000 m/sec, respectively).

If we approximate the sequence as a layer with constant wave speed, we may consider that head waves are being studied for two types of simplified models, as shown in Fig. 3. The properties of the single layers and sequences of layers are listed in Table 1. The index 2 designates the average property for a sequence of layers, while the indices 1 and 3 designate the average properties (over a thickness of about one wavelength) of the overlying and underlying rocks, respectively. A horizontal bar denotes the average properties characterizing the overlying beds (complete section).

1 in Fig. 3a, and 2 and 3 in Fig. 3b correspond to a sequence of layers having a small total thickness (6-25 m), and $l/\lambda \leq 0.30$.

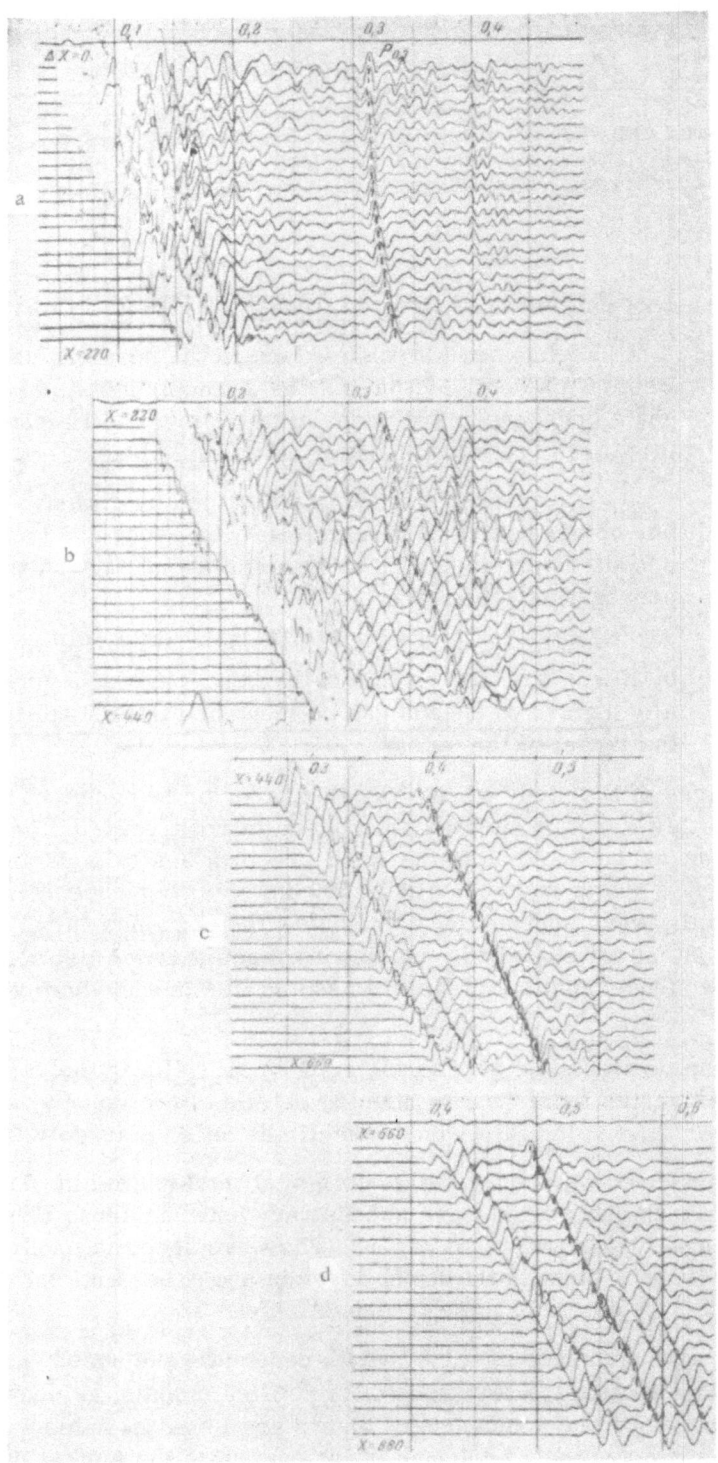

Fig. 4. Seismograms showing the longitudinal wave $P_{0.3}$ reflected from a layer with h = 280 m. Station IV, shot point 1420, with geophones spaced from 0–880 m.

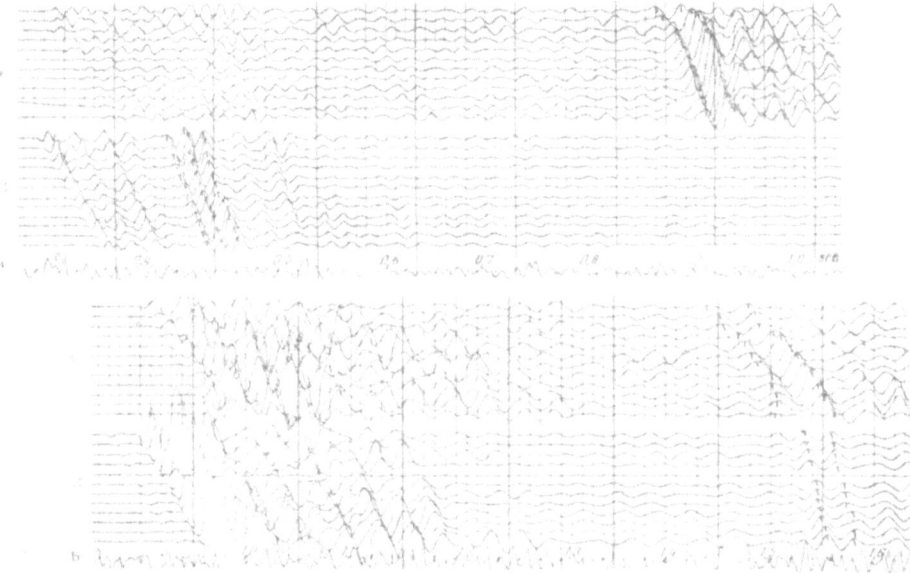

Fig. 5. Seismograms showing PP and PS reflected waves from a layer with h = 280 m. a) Offset 400-510; b) offset 1440-1550. PP waves over the time interval 0.4-0.5 sec (a) and 1.0-1.1 sec (b). PS waves over the time interval 0.8-1.0 sec (a) and 1.4-1.6 sec (b).

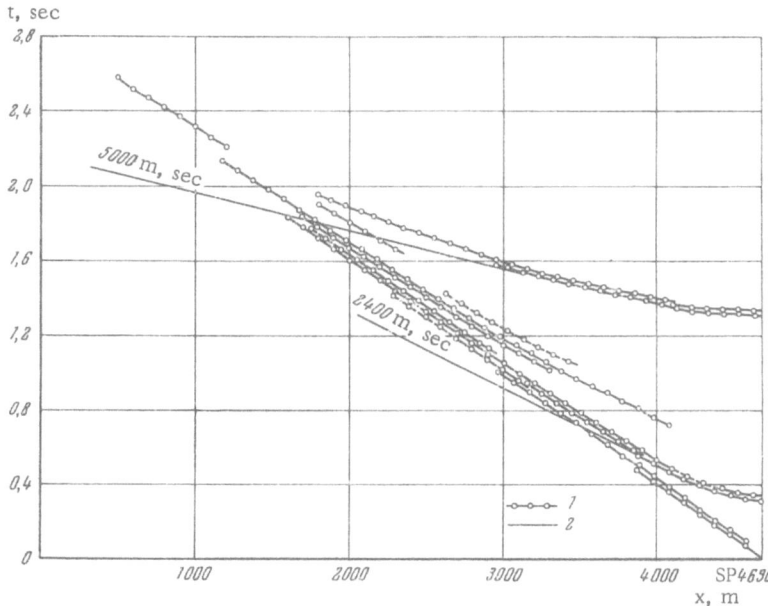

Fig. 6. Travel times for longitudinal waves. 1) Experimental travel times; 2) theoretical travel times for head waves from a layer with h = 280 m (V_h = 2400 m/sec) and from a layer with h = 1300 m (V_h = 5000 m/sec).

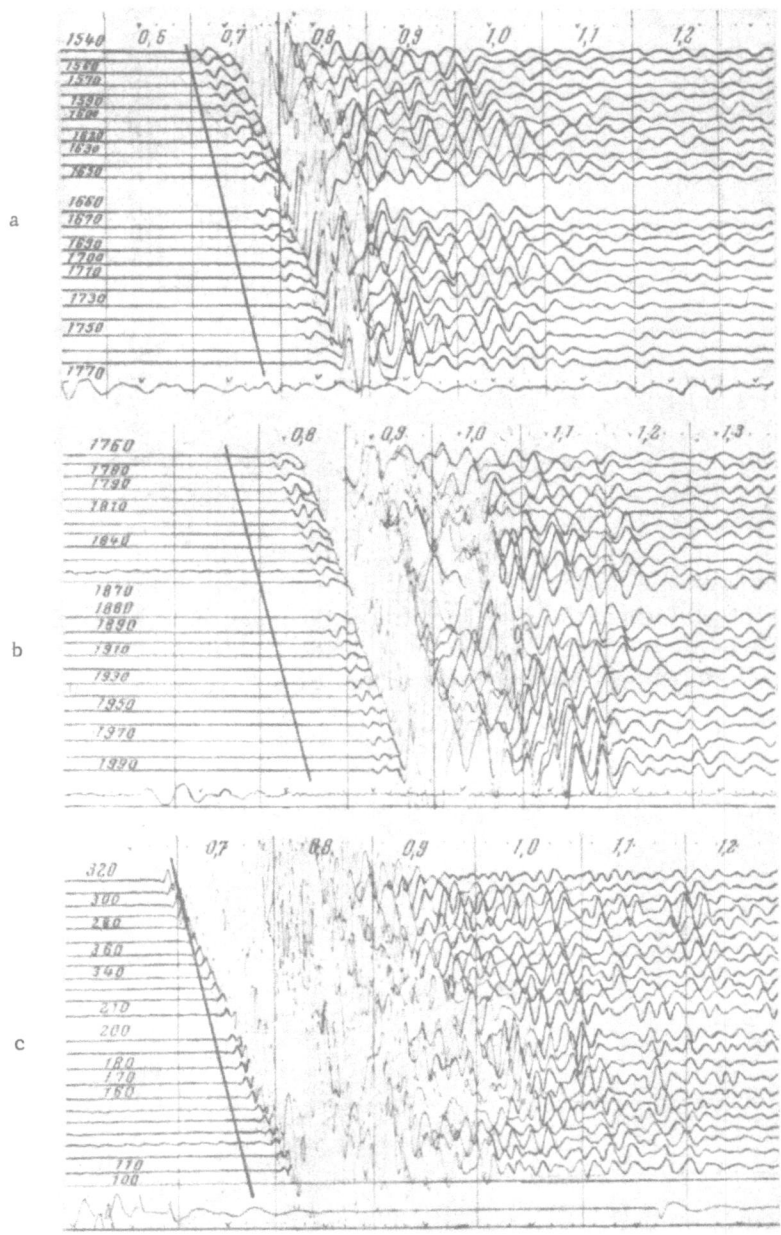

Fig. 7. Seismograms showing first arrivals. The lines denote
the arrival times for head waves (h = 280 m, V_h = 2400 m/sec);
a and b are broadband recordings, and c was made with filter
settings 8-10. The distances from the shot point are 1080–
1300 m (a), 1300–1520 m (b), and 1000–1220 m (c).

The thickness of the beds, their degree of inhomogeneity, the wave speeds in the rocks above and below, the discontinuity in wave speeds at the boundaries, and the depths to the beds all differed between the two target horizons.

In the survey area, the general character of the velocity section (values for formation wave speeds in the overlying section, and the presence of beds with high wave speeds) was consistent over large areas. The velocity profiles sometimes differ in detail. In order to compare wave characteristics with geology, sections were drawn by correlating between wells or by extrapolating the section found in one well over larger or smaller areas. The uniformity of the section was judged by consistency of the reflected wave properties recorded close to the shot points. There may be some chance for error involved in this.

Head Waves from a Thin Layer in a Homogeneous Medium

The target of the study was the thin layer at a depth of about 280 m. Surveys were carried out in three areas, where there was some variation in bed thickness (6-11 m) and in the ratio of wave speeds VP_1/VP_2 (0.69-0.82).

Head waves from this bed were recorded in none of the areas. At the same time, the reflected PP and PS waves had large amplitudes and could be traced to offset distances of 5-10 h where the arrival angles were equal to or larger than the critical angle (Figs. 4, 5). Over part of the interval over which it can be traced, the reflected wave forms the apparent first arrival.

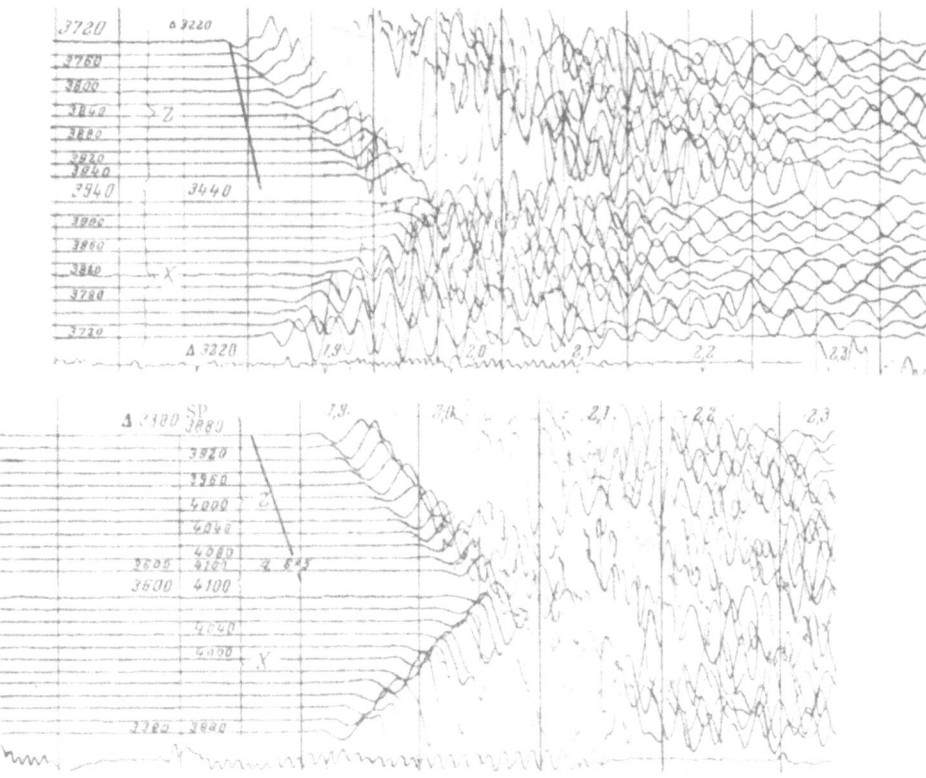

Fig. 8. Seismograms showing first arrivals. The lines indicate the calculated arrival times for head waves (h = 1300 m, V_h = 5000 m/sec).

Figure 6 shows the travel—time curves for the observed data; the light lines indicate the theoretical travel times for head waves.

Figure 7 shows seismograms obtained in areas where calculations indicate that the first arrival must be a head wave. The waves are not apparent even on records made with high amplification with broadband equipment. Neither are they recognizable on filtered records, including those made with high-frequency equipment (Fig. 7c).

We carried out the following efforts with the purpose of obtaining larger amplitudes and higher-frequency recordings:

a) Larger shots were used; charge sizes up to 30 kg were used with offset distances of 1-2 km (such offset distances would cause any head waves to be registered first); the shot depth was 50 m;

b) shots were detonated inside tubing in order to enhance the high-frequency part of the spectrum; and

c) seismic vibrations were recorded in drill holes beneath the low-speed layer in order to avoid the filtering action of the low-speed layer.

Despite all of this, no head waves were recorded.

Based on the amplitudes of the reflected waves and of the noise in the areas where head waves should have been recorded, we conclude that the ratio A_{refl}/A_{head}, of the amplitude of the longitudinal reflected wave to that of the head wave is greater than 200 for $x/2h \geqq 2$.

No significant change in the dominant frequency on the records was noted either at the Earth's surface or below the low-speed layer when the shot conditions were changed. The dominant frequency of the waves reflected from this layer was 50-60 cps.

The ratio of layer thickness to wavelength was determined for the dominant frequencies on the records, and the maximum frequency in the reflected wave spectrum ($f = 50$-60 cps). The variation of l/λ_2 for various frequencies and various layer thicknesses lies in the limits 0.12-0.32; the ratio of the depths h of the layer to the wavelength in the overlying beds λ_1 is about 8-10.

According to theory, for a thin layer with a ratio of depth of burial to wavelength h/λ, wave speed contrast V_{P_1}/V_{P_2} and ratio of offset distance to burial depth x/h the same as under the experimental conditions ($h/\lambda_1 = 10$, $V_{P_1}/V_{P_2} = 0.6$, and $x/2h = 2$-3), and with no attenuation in the medium, the ratio A_{refl}/A_{head} is about 20. With attenuation, this ratio may be somewhat less. It may be assumed that the small thickness of the refracting layer is responsible for the few, relatively intense, head waves that were formed ($A_{refl}/A_{head} > 200$).

Head Waves from a Thin Layer in an Inhomogeneous Medium

The target of this study was the layer with a high wave speed ($V_2 \approx 5000$ m/sec) at a depth of 1260-1300 m. The surveys were carried out in two areas. In one area, the layer thickness was 13 m, and in the other, 25 m; the wave speed in the layer differed slightly (approximately by 500 m/sec). Beneath the layer, the formation wave speed was 3600 m/sec in the first case, and 3000 m/sec in the second case, while the wave speed in the overlying rocks was 2500 m/sec.

According to calculations, the head waves from this layer should be recorded as a first arrival beginning at an offset distance of about 3 km, and the abscissa of the depth intercept should be about 950 m.

No head waves were recorded from this layer. The PP reflected waves had large amplitudes and could be traced at emergence angles ranging from less than critical to more than critical.

The observed and theoretical travel times are shown in Fig. 6; seismic records obtained at offset distances at which head waves from a layer with a wave speed of 5000 m/sec should be recorded first are shown in Fig. 8. Thus, head waves were not recorded for these thin layers in an inhomogeneous medium.

The ratio of bed thickness to wavelength is about 0.13-0.16 in one area, and 0.20-0.24 in the other ($V_{P_2} \approx 5000$ m/sec, $l_1 = 13$ m, and $l_2 = 25$ m; the frequency is taken as the dominant frequency recorded broadband, with $f = 40$-50 cps).

The values for l/λ_2 for this layer are less reliable than in the case of the layer in a homogeneous medium, inasmuch as thicknesses observed in wells were extrapolated by distances up to 3-4 km.

The data which were obtained allow us to take note of the general characteristics for two types of high-speed layers: a) the high amplitudes of reflected waves for emergence angles ranging from less than critical to more than critical; and b) the absence of head waves on the record.

We cannot exclude the possibility that for some special set of conditions, head waves might be recorded (very large charges, concentration of the shot energy at high frequencies from small shots, and so on), but it is obvious that the relative amplitudes will be very small.

Comparison of These Data with Data of Other Investigators

A brief resumé of field, model, and theoretical work on the relationship of refracted wave forms to layer thickness was included in the introduction to this paper. We will compare these data with our own in the following paragraphs.

Comparison of the Data with Model Data and Theoretical Studies. Our data agrees quantitatively with the modelling data for a uniform thin layer in a solid medium [8-10]. Agreement can be seen in the following behavior: 1) the low intensity of head waves for small l/λ_2; and 2) the high intensity of PP and PS waves in comparison with head waves.

The values for l/λ_2 at which head waves could not be identified on the field records were higher than those for the models. This discrepancy may possibly be explained by the effects of attenuation and different effects from scattering at different values for h/λ_1 (in the field experiments, $h/\lambda_1 = 8$-26, while in the model studies described in reference [10], $h/\lambda_1 = 2.2$).

The results which we have obtained do not agree with data from liquid/solid models [6]. According to reference [6], the head waves from thin layers in a liquid medium may be nearly as intense as those from a thick layer, and decay slowly with distance.

The lack of head waves from thin layers is in qualitative agreement with the results of theoretical work [13] which indicates the absence of low-frequency head waves in the case of a thin layer in welded contact with the surrounding medium. In the light of the theoretical and model studies, we may conclude that the absence of head waves in the field studies is related to their low intensity for low values of l/λ_2.

Consideration of Field Data from References [3, 4]. No acoustic logging data were shown for the surveys described in references [3, 4]. In view of the lack of adequate data on the thickness of the refracting layer and the distribution of wave speeds in the layer, there is some doubt about the reliability of the values for l/λ_2 given in reference [3]. Moreover, the experimental conditions in reference [3] were quite special; the refracting layer was at shallow depth and the value for h/λ_1 was less than 1. It is possible that in this case the intensity of the head waves may be larger than in the case of larger h/λ_1. Theoretical calculations for the boundaries of thick layers indicate that the relative intensity of head waves increases in comparison with that of reflected waves with decreasing h/λ_1 [14].

In the example described in reference [4], the refracting layer was taken to be a layer of gypsiferous anhydrite in a section of sandstone and shale beds, with $h/\lambda_1 \approx 10$. These experimental conditions were close for examples 1 and 2 of the present paper, but head waves were recorded ($l/\lambda_2 \approx 0.15$). The relative intensity of the reflected waves in reference [4] was less than those in examples 1 and 2 of the present paper. This difference could hardly be associated with a difference in the wave speed contrast (0.5 in the fore-Carpathians, 0.7 for the layer with h = 280 m, and 0.5 for the layer with h = 1260-1300 m). An explanation of the cause of this difference is difficult to find. The transverse wave speeds have not been studied in either case. The thickness of the refracting layer may possibly be given incorrectly in reference [4]. Moreover, in both regions the layer is inhomogeneous, and the effect of inhomogeneity on the form of head waves has not been studied.

Conclusions

We have tried to record head waves from a thin layer in an actual medium where the velocity profile is known in detail.

1. For a reasonably homogeneous layer embedded in a practically homogeneous medium (depth of 280 m, h/λ_1 = 8-10), intense longitudinal and shear reflected waves (PP and PS) were recorded and could be traced to offset distances of 6-10 h; that is, for emergence angles ranging from less than critical to more than critical. Head waves were not recorded from the thin layers with $l/\lambda_2 \lesssim 0.3$, which is evidence of their low relative intensity.

2. Intense reflected waves were recorded at emergence angles ranging from less than critical to more than critical (traceable to offset distances of 2-2.5 h) for a thin layer in an inhomogeneous medium (the wave speeds in the beds above and below the layer were less than the wave speed in the layer) for h = 1300 m, h/λ_1 = 20-50, and $l/\lambda_2 \lesssim 0.2$. Head waves were not recorded for $l/\lambda_2 \lesssim 0.2$.

3. The data which have been obtained lead us to suppose that it is possible to determine the nature of a layer on the basis of the wave forms of various types. Thus, the presence of high-amplitude PP and PS waves traceable to large distances and the lack of or low amplitude of head waves may possibly be criteria indicating a thin reflecting layer.

In view of these data, we must review the basic concepts on which refraction surveying is based.

In refraction surveys it is assumed that the head waves in sedimentary rocks are frequently related to thin high-speed layers, and it is considered that the layer thickness may be very small in comparison with a wavelength [1-4].

The data we have obtained indicates that head waves from thin layers will be so weak that they cannot be recorded.

It will be necessary to carry out experiments specifically designed to test the generality of our results about the low relative amplitudes of head waves, as well as on the properties of the medium which lead to the development of head waves with large enough amplitudes that they may be identified on records. The recording of a wave as a first arrival is rarely one of the criteria for identifying a refracted wave type (either head wave, or a simply refracted wave). The data presented here, as well as the results of others [15 and others] indicates that the first arrival may be a reflected wave. They may be considered to be the first arrival because of their large amplitude in comparison with the head waves.

Thus, recording a wave as a first arrival is not a sufficient criterion for determining the nature of the wave.

It may be assumed that an error was made in identifying wave types in some of the earlier studies. Waves which have been assumed to be head waves (from thin layers, from boundaries with weak contrasts in wave speed, from screening layers, and so on) may actually have been reflections at large emergence angles or true refractions. In general, we need to reexamine the role of head waves in forming a seismic record.

All of these troublesome questions require broad study, directed at working out the physical basis for the refraction method. It is not inconceivable that work in this direction would permit broadening the applicability of the method as well as its precision.

LITERATURE CITED

1. G. A. Gamburtsev, Yu. V. Riznichenko, I. S. Berzon, A. M. Epinat'eva, I P. Pasechnik, I. P. Kosminskaya, and E. V. Karus, Correlation method for refracted waves, in: Handbook for Seismic Exploration Engineers, Akad. Nauk SSSR, 1952.
2. G. A Gamburtsev, Yu. V. Riznichenko, M. K. Polshkov, E. V. Karus, A. M. Epinat'eva, and I. P. Kosminskaya, Correlation of refracted waves in the Eastern Apshiron, Dokl. Akad. Nauk SSSR, No. 50 (1945).
3. I. S. Berzon and A. M. Epinat'eva, On seismic screening, Izv. Akad. Nauk SSSR, Ser. Geogr. i Geofiz., No. 6 (1950).
4. I. S. Berzon, A. M. Epinat'eva, G. N. Pariiskaya, and S. P. Starodubrovskaya, Wave forms of seismic waves in real media, Akad. Nauk SSSR, 1962.
5. Yu. V. Riznichenko, On scattering and attenuation of seismic waves, Tr. Geofiz. Inst. Akad. Nauk SSSR, No. 35:162 (1956).
6. N. I. Davidova, Model study of the relation of wave forms for longitudinal head waves to the thickness of a refracting layer, Izv. Akad. Nauk SSSR, Ser. Geofiz., No. 1 (1962).
7. O. G. Shamina, Decay of head waves from thin layers for liquid or sliding contacts, Izv. Akad. Nauk SSSR, Ser. Fiz. Zemli, No. 3 (1965).
8. M. Lavergne, Étude sur modèle ultra du problème des couches mines en séismique réfraction, Geophys. Prospecting, No. 1 (1961).
9. F. K. Levin and J. Ingram, Head waves from a bed of finite thickness, Geophysics, No. 6 (1962).
10. P. G. Gil'berstein and I. I. Gurvich, Study of head waves from layers of different thickness on two dimensional models, Izv. Akad. Nauk SSSR, Ser. Geofiz., No. 11 (1963).
11. Yu. V. Riznichenko and O. G. Shamina, On elastic waves in solid media based on investigations in two dimensional models, Izv. Akad. Nauk SSSR, Ser. Geofiz., No. 7 (1957).
12. R. Y. Donato, Amplitude of P-head waves, J. Acoust. Soc. Am., 36(1) (1964).
13. L. A. Molotkov and P. V. Krauklis, On the formation of low-frequency head waves in thin layers, Izv. Akad. Nauk SSSR, Ser. Geofiz., No. 6 (1963).
14. A. M. Epinat'eva, Study of longitudinal seismic waves which propagate in an actual layered medium, Tr. Inst. Fiz. Zemli, Akad. Nauk SSSR, No. 14 (1961).
15. A. G. Averbukh, M. M. Gorvach, and É. P. Sumerina, On the physical nature of waves recorded as first arrivals in the refraction method, Prikl. Geofiz., No. 36 (1965).

SHEAR WAVES REFLECTED FROM A THIN HIGH-SPEED LAYER

I. S. Berzon, L. I. Ratnikova, and V. A. Mitronova

Introduction

The principal targets in our studies on suitable representations for real media, which have been carried out in the Krasnodar Belt, are two thin layers situated at depths of 280 and 1350 m. Strong shear wave reflections from these boundaries were recorded. The present paper gives the results of an interpretation of data on shear wave reflections and the comparison of these results with some theoretical calculations. The principal attention has been paid to the PS wave reflected from the boundary at 280-m depth (t_1PS), inasmuch as quite good data were obtained for it. We have studied the PS wave from the layer at 1350-m depth (t_2PS) to a lesser extent.

The measurements were made with horizontal geophones and PMZ-model equipment; an SPM-16 seismograph was used. All field recordings were made with broadband filters, and filtering and mixing were done at a replay center.

Record Character

Recognition Ranges for Waves. The wave t_1PS, reflected from the thin high-speed layer at 280-m depth, was traced with single geophones from directly over the shot point to offset distances of 5 or 6 times the depth of burial, for small shots ($Q = 200-600$ g) (see Fig. 1).

The wave t_2PS was not recorded near the shotpoint, as might be expected; it could be recognized on the records at a distance of 400-600 m, but the amplitude was about the same as that of the noise. Wave phases could be correlated continuously starting at a distance of 1000 m, and could be carried to the largest spreads used (see Fig. 2). On some profiles, the uncorrelated noise level was large also for x > 1000 m. In these cases, the t_2PS waves were recognized on replay records made using low-pass filtering ($f_{max} = 10-13$ cps) and mixing of 4 or 7 channels with time shifting.

The t_1PS wave has large amplitude on the records, has a sharp, clear form and has an apparent frequency of about 25-30 cps. The t_2PS wave is significantly larger than the surrounding noise for offset distances x > 1000 m and has an apparent frequency of 20 cps. The wave form is inconsistent, and the number of phases and the relationships between them vary.

Travel Times for the Recorded Waves. Examples of the travel−time curves for the waves t_1PS and t_2PS are shown in Fig. 3. The travel−time curves for the t_1PS are long and curved (constructed over the interval from x/h = 0 to x/h = 5-6) and have relatively small scatter of the points about an average curve. The travel times for the t_2PS waves show appreciable scatter of the points about an average curve, and the parts of the curves close to the shot point are missing.

Fig. 1. Examples of records of the wave t_1PS obtained at distances of 0-2 km from the shot point. Charge size, 0.4 kg; recorded broadband.

Interpretation of Travel Times for the t_1PS and t_2PS Waves

All of the travel times were interpreted using the computer technique described in reference [1]. Two basic problems must be resolved in interpretation: 1) to establish the uniqueness or nonuniqueness in the use of the various segments of the travel−time curves for the solution of the inverse problem; and 2) to find the optimum interval over which the interpretation of the travel−time curves will provide results which best coincide with well control.

Interpretation of the t_1PS Travel Times. These travel−time curves were symmetrical about the shot point. The intervals selected from the travel−time curves and the values for V_P, V_S, and H which were obtained are listed in Table 1. It is seen that in interpreting travel−time segments, for $l/2 < 2H$ (where l is the total length of the travel−time curve), the results which are obtained differ significantly from the correct values − the speed for longitudinal waves is too high, and the speed for shear waves and the depth are too low. This is apparently related to the fact that at large angles of incidence of the waves at the boundary, the medium cannot be considered to be uniform. The presence of minor layering or weak velocity

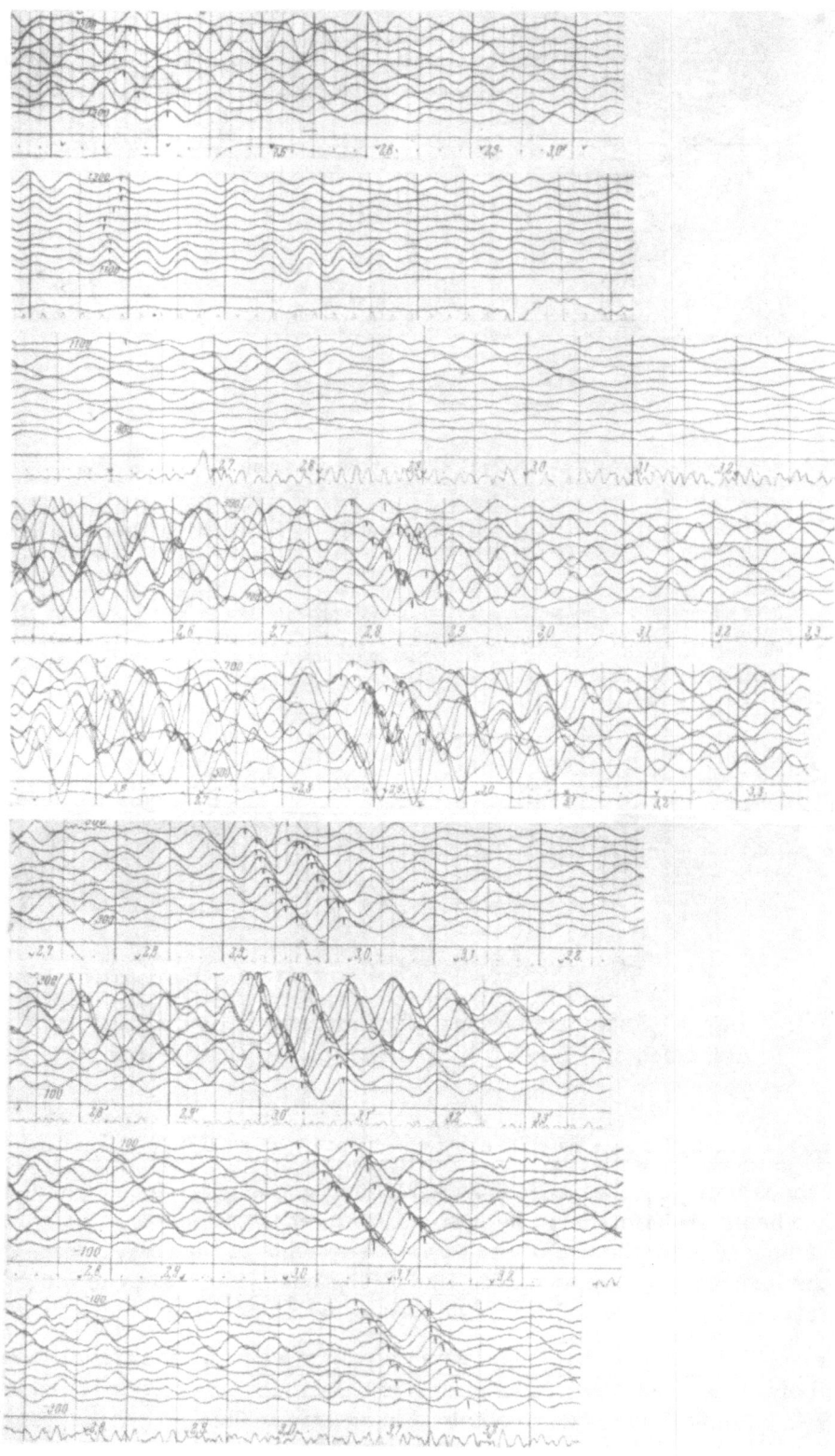

Fig. 2. Examples of records of the wave t_{2PS} obtained at distances of 200–2000 m from the shot point. Recorded broadband.

TABLE 1

Wave	Travel-time curve segment, m	V_P, m/sec	V_S, m/sec	V_S/V_P	H, m	Data from acoustic logs and downhole surveys		
						V_P, m/sec	V_S, m/sec	H, m
t_{1PS}	-200 $+2000$	1790	300	0.17	200			
t_{1PS}	-600 $+580$	1580	470	0.3	280			
t_{1PS}	-1300 $+1240$	1850	340	0.18	221	1640	430	270
t_{1PS}	-600 $+600$	1630	440	0.27	270			
t_{2PS}	400—1000	2000	730	0.36	1560			
t_{2PS}	-370—980	2000	850	0.42	1330	2000	—	1350
t_{2PS}	980—1000	2680	730	0.27	1650			
t_{2PS}	500—1300	2000	730	0.36	1360			

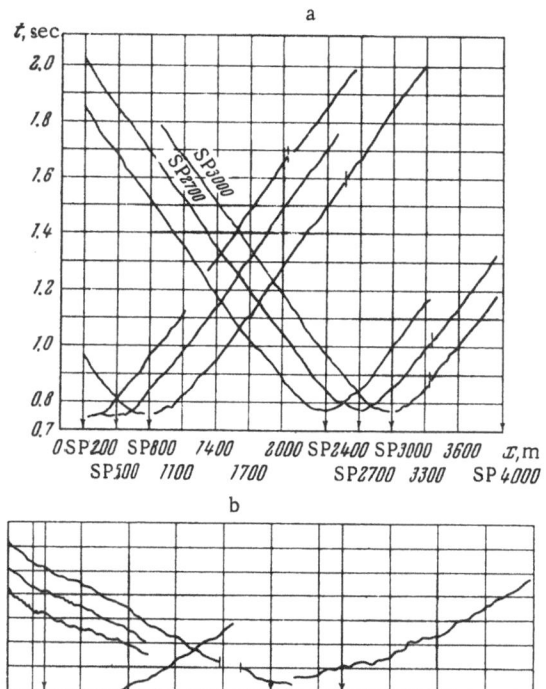

Fig. 3. Travel—time curves for recorded waves. a) t_{1PS};
b) t_{2PS}; c) profile IV-4.

gradients in the section distorts the form of the travel—time curve. Therefore, in approximating the observed travel—time curves with a theory based on the assumption of a uniform overburden, the best match gives parameters which differ from the actual values.

In interpreting travel—time curves with $l/2 \sim 2H$, the results which are obtained agree well with acoustic logging data. The use of short segments of the travel—time curves with $l/2 < H$, as one might expect, gives poorer results than the use of segments with $l/2 \approx 2H$. This is caused by the larger scatter of the points on the travel—time curves close to the shot point, where the waves are of low amplitude, and irregular noise is present.

TABLE 2

	Wave speed in the overlying medium, m/sec		Wave speed in the layer, m/sec		Wave speed in the underlying medium m/sec		Layer thickness, m
	v_P	v_S	v_P	v_S	v_P	v_S	
Halfspace	1880	450	2500	1000—710			—
Uniform thin layer	1880	450	2500	1000—710	1880	450	7
			1700	370	1880	450	4
Sequence of two uniform thin layers	1880	450	2500	1000—710			7

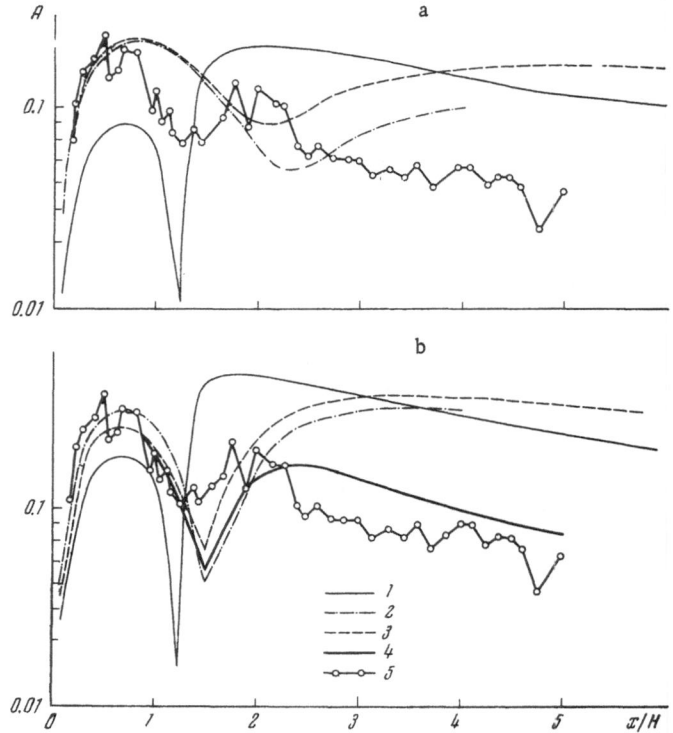

Fig. 4. Observed and calculated curves for the relationship between t_1PS amplitude and distance. a) $V_{S_2} = 710$ m/sec; b) $V_{S_2} = 1000$ m/sec; 1) halfspace; 2) uniform layer; 3) inhomogeneous layer; 4) inhomogeneous layer with consideration of scattering and attenuation in the overlying rocks; 5) experimental.

Interpretation of the travel times for the t_1PS waves gives an average speed for longitudinal waves in the medium as 1580-1620 m/sec, a speed for the shear waves of 440-470 m/sec, and a depth to the reflecting surface of 270-280 m. Thus, the reflecting horizon giving rise to the strong PS wave can be identified as the thin high-speed layer with a high degree of confidence.

Interpretation of the t_2PS Travel — Time Curves. Interpretations were made by two methods: 1) a method of successive approximations with the use of effective parameters

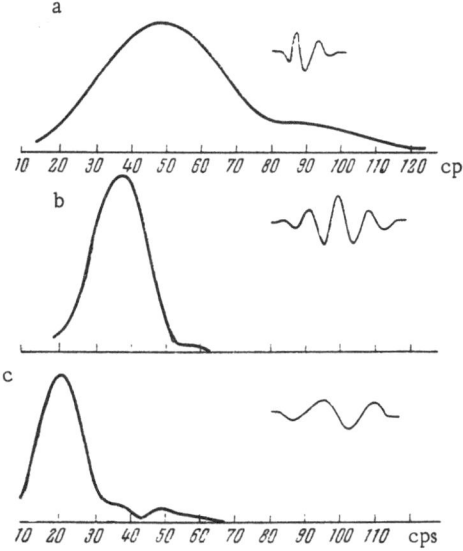

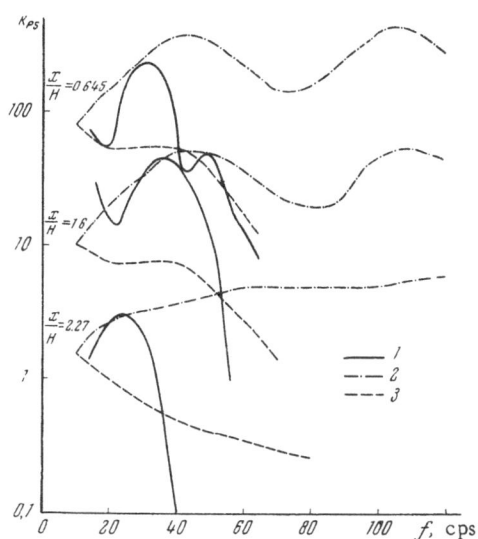

Fig. 5. Spectrums of recorded waves. a) Direct wave, profile IV-5, shot point (−300), broadband, b) t_1PS wave, profile IV-5, $\Delta = 180$ m; c) t_2PS wave, profile IV-5, $\Delta = 820$ m.

Fig. 6. Response functions of the layer for the t_1PS wave. 1) Observed; 2) theoretical, without consideration of attenuation; 3) theoretical, with consideration of attenuation.

and a curve for the variation of longitudinal wave speed with depth, obtained from acoustic logs [3]; and 2) a computer method as described in reference [1]. The results are listed in Table 1. As is apparent from these data, there is considerable scatter. The values for parameters obtained in the interpretation differ from the well control by 15-20%. The low accuracy of the computations is caused by the lack of the curved portion of the travel−time curve, which makes the inverse problem nonunique, and by the large scatter of points on the travel−time plot, as well as by the difficulty in making corrections to travel−time curves for low-frequency waves, especially the correction for phase shift. Therefore, no reliable conclusions concerning the boundary from which the t_2PS waves are reflected can be drawn from an interpretation of the t_2PS travel times. However, vertical seismic profiling with a three-component detector established that this boundary is that of the thin layer at 1350-m depth [2].

We note that even for a comparatively great depth to the reflecting boundary ($H \approx 1350$ m), the value for V_S/V_P is about 0.36. It is necessary to know this in wave form calculations for various classes of waves for comparison with experimental data obtained in a thick sequence of clastic sediments.

Amplitudes of the Recorded Waves

The positive correlation of t_1PS waves over long intervals, their simple form and dominant amplitudes allow us to construct an amplitude−distance curve with a good degree of reliability. An example of such a curve obtained by averaging three sets of data is shown in Fig. 4. The first maximum in amplitude is situated at distances of 170-200 m ($x/H = 0.61-0.71$), a minimum in amplitude is situated at distances of 360-420 m ($x/H = 1.3-1.5$), and a second maximum is situated at distances of 560-600 m ($x/H = 2-2.1$). A theoretical curve for changes in amplitude for reflection from a thin layer with properties as listed in Table 2 (for a frequency of 25 cps) is shown for comparison, as well as a curve for reflection from the surface of a halfspace with the same wave speed parameters.

TABLE 3

Well	Wave speed in the overlying rocks, m/sec		Wave speed in the layer m/sec		Wave speed in the underlying rocks, m/sec		Layer thick-ness, m
	v_P	v_S	v_P	v_S	v_P	v_S	
52	2470	825	5110	2550	3000	1000	24
39 (1 layer)	2800	750	3320	1100	2940	830	50.5
39 (14 layers)	2800	750	2100	525			8.5
			2900	830	2940	1830	2.5
			5650	2820			1.0
			2740	780			1.0
			5000	2500			3.0
			3780	1550			2.5
			2760	790			2.5
			5250	2640			2.5
			2280	650			4.0
			3880	1550			1.5
			4680	2340			9.0
			4160	1660			3.0
			5220	2610			4.0
			3480	1390			5.5

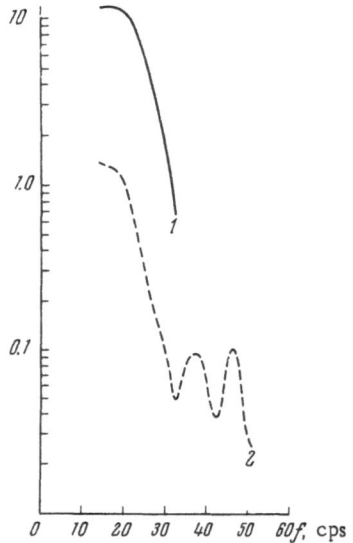

Fig. 7. Experimental response functions for a layer computed from spectrums for the t_2PS wave at two points along a profile ($x/H = 1.22$).

The values for longitudinal wave speed and thickness were taken from acoustic logging data described by Karus and Saks in this collection of papers, and the value for shear wave speed in the overlying rocks was obtained from a surface/downhole survey in a drill hole. The value for V_{2S} was not determined, and so, in the calculations, various values were used within the limits listed in Table 2.

As is apparent from the curves which are shown, the change in reflection factor for the halfspace is exactly the same for both cases which were considered. The curves for the variation of the modulus of the reflection factor from a thin layer differ markedly for different values of V_2S. The minimum values for amplitude in the case where $V_{2S} = 710$ m/sec is located in the region $x/H = 2.1-2.3$, while for $V_{2S} = 1000$ m/sec, it is located at $x/H = 1.5$. The curves for the reflection factor for the models of uniform and inhomogeneous thin layers (with a fixed maximum wave speed V_{2S} in the layer) which were considered are quite similar and differ only in the ratios of levels over the initial and terminal regions. The observed amplitude curves agree best with the theoretical curve computed for a thin layer with $V_{2S} = 1000$ m/sec. The transition from curves for the change in the reflection factor to curves for amplitude with consideration of scattering and attenuation gives similar results for the theoretical and observed curves over all intervals (Fig. 4b). The value for the attenuation factor for longitudinal waves at 25 cps was taken from Kuznetsov and Gamburtsev (next article), and was $\alpha_P = 0.006$/m in the low-speed layer, and $\alpha_P = 0.0004$/m in the rocks beneath the low-speed layer. The value for the attenuation factor for shear waves was computed from the value for α_P, using the following ratios for attenuation decrements [3]: for the low-speed layer, $\Delta_P/\Delta_S = 1.0$, and for the underlying rocks, $\Delta_P/\Delta_S = 0.25-0.3$. The values obtained for α_S are 0.012/m for the low-speed layer, and 0.005/m in the underlying rocks.

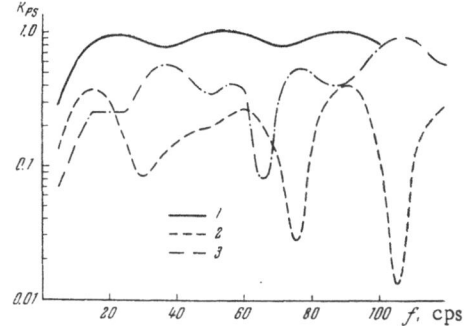

Fig. 8. Theoretical response functions computed for the layer at a depth of 1350 m, recognized from acoustic log data.

TABLE 4

v_P, m/sec	v_P / v_S
3000	⩾3.0
3000—4000	3.0—2.5
4000—5000	2.5—2.0

The curves for the amplitude of the t_2PS waves have been constructed only for distances starting at 1000 m, and decrease monotonically with distance. Inasmuch as the attenuation for the medium at greater depths was not determined, no comparison was made with the results of the theoretical calculations.

Spectrums of the Recorded Waves

An attempt was made to compare the observed spectrums for the t_1PS and t_2PS waves with spectrums computed from theoretical considerations [4].

The observed spectrums were computed as amplitude spectrums for the reflected t_1PS wave (Fig. 5) and the direct longitudinal wave recorded at the surface close to the shot point. The observed response function for the t_1PS wave and that computed theoretically for the third variant of function for the layer, and the spectrum of attenuation in the overlying rocks is also shown on this figure. We used the relationship between α_P and frequency given by Kuznetsov and Gamburtsev [this volume], and calculated the relationship between α_S and frequency from this. The observed and theoretical response functions computed without attenuation in the overlying rocks for small values of x/H (≤ 0.645) are underdamped in form. However, the range of the maximum in the observed response function is much narrower than that of the theoretical function. As is evident from the curves which are shown, lack of consideration of attenuation in the overlying rocks cannot explain this discrepancy. A still stronger difference is noted at x/H = 2.27. In this case, the observed response function still exhibits an underdamped form while the theoretical curve increases montonically with frequency. When attenuation is considered, the curve decreases monotonically with frequency.

The cause for these discrepancies are not readily apparent; possible causes may be as follows:

1. The response function is very sensitive to the parameters for the layer, particularly to the thickness of the layer and the transverse wave speed. The layer thickness was selected on the basis of data on longitudinal wave speeds. It is possible that the change in shear wave speed in the medium is different, and that a larger thickness of rock is involved in generating the reflection than was assumed in the calculations.

2. We selected the wave recorded at the surface close to the shot point as the direct wave, in computing the observed response functions. In so doing, we did not consider the difference in filtering action of the low-speed layer on longitudinal and shear waves.

We also obtained response functions for the t_2PS wave (Fig. 7) which showed a persistent form at various points along the survey profiles. The profile along which these measurements were made passed through two wells in which acoustic logs had been run. According to these logs, the layer could be considered uniform at one of the wells, but had to be divided into a number of thin layers at the other, with properties as listed in Table 3.

The longitudinal wave speed and layer thickness were taken from the acoustic logs. The shear wave speeds were calculated from relationships (Table 4) obtained experimentally in reference [1].

The response functions for this layer computed for the PS wave with the parameters listed above are shown in Fig. 8. It is apparent that the response functions for the uniform and inhomogeneous layers differ markedly even at low frequencies, $f \leq 40$ cps. The response functions for uniform layers, recognized in various wells, also show marked differences. All of them have in common a clear maximum at low frequencies, $f = 15$-20 cps, which is in agreement with the observed functions (Fig. 7).

The study of spectral response functions for sequences of thin layers for waves of various types and with inclined incidence, and their comparison with theory has only begun. A volume of such experimental and theoretical data would permit a determination of the effects of various parameters for the medium on the form of the response function and might point the way to their use for the solution of the inverse problem.

Conclusions

1. The highest amplitude shear-wave reflections were related to a thin layer in the section.

2. The travel−time curve interval giving the interpretation which best agrees with acoustic well log data was determined.

3. The nature of the change in amplitude of the PS wave with distance was established, and good agreement with the theoretical behavior for a thin-layered medium was found.

4. It was pointed out that the PS response function for a layer is sensitive to variations in the wave speed characteristics obtained from acoustic logs.

5. It was pointed out that there are significant differences between the observed response functions and the theoretical ones; further studies will be required to explain this.

LITERATURE CITED

1. I. S. Berzon, L. I. Ratnikova, and M. I. Rats-Khizgiya, Shear Reflected Waves, Izd. Akad. Nauk SSSR, 1966.
2. E. I. Gal'perin and A. V. Frolova, Study of shear waves in vertical seismic profiling, Izv. Akad. Nauk SSSR, Ser. Fiz. Zemli, No. 9 (1966).
3. Yu. I. Vasil'ev and G. I. Gurevich, On the relation between the decrement of attenuation and the propagation speeds for longitudinal and shear waves, Izv. Akad. Nauk SSSR, Ser. Geofiz. No. 12 (1962).
4. L. I. Ratnikova and A. L. Levshin, Calculation of response functions for thin-layered media, Izv. Akad. Nauk SSSR, Ser. Fiz. Zemli, No. 2 (1967).

METHOD FOR RECORDING DIRECT
LONGITUDINAL WAVES AT THE SURFACE
AND WITHIN A MEDIUM

V. V. Kuznetsov and A. G. Gamburtsev

At present, in the solution of a variety of problems in seismic exploration, there is a need for detailed data on the properties of actual media and the nature of propagation of seismic waves in such media. The wave forms and arrival times of waves recorded both at the surface and at interior points in the medium are used in obtaining these data [1-6].

Of all the wave types, direct waves are the most important. These waves contain the major part of the information about the medium because they are recorded early on the seismogram where the noise level is relatively low and the waves contain high frequencies and a broad spectrum.

Some ideas about methods for recording direct waves at the surface and in shallow and deep drill holes have been described in the literature [1-5, 7-9]. However, up to the present time, there has been no adequate discussion of the specific problems involved in recording seismic waves both at the surface and within a medium: These specific problems are as follows:

1. How close the recording point may be to the shot point with asymptotic expressions remaining valid [10] (commonly, the direct waves are recorded at distances of no more than 100-150 m from the shot point);

2. the necessity of selecting an offset distance from the mouth of a drill hole to the receiver so that the disturbed zone about the drill hole has no effect on the form of the direct wave;

3. the necessity of selecting an offset distance from the shot point to the receiver so that there will be no interference between the direct and other waves; and

4. the best method for implanting a geophone at a point within the medium to record the direct waves.

In this paper, we propose a method for recording direct waves based on an analysis of experimental data. The method was developed for the following conditions:

1. Measurements were made with normal or near normal incidence;

2. the direct waves were recorded with relatively small charge sizes (from one squib to 1.2 kg), so that secondary impulses were not present on the records;

3. measurements were made outside the region of inelastic deformation, over a depth range of 0-100 m.

Equipment

Two types of equipment were used in making the measurements; broadband recording equipment with provisions for integrating and differentiating, and equipment provided with magnetic tape recording. With this equipment, recordings were made broadband in the field, and filtering was done later at the replay center.

Electrodynamic geophones with a frequency of 32 cps and galvanometers with frequencies of 130 and 250 cps were used.

In the development of methods for recording direct waves, geophones with large mechanical motions were used. This has led to the necessity for avoiding possible distortion of the seismic signal from coil travel exceeding the linear range for the standard SPM-16 geophones. During the study it was established that for charge sizes up to 10 kg at depths of 10-15 m, the direct wave could be recorded at the mouth of the shot hole with an SPM-16 geophone without distortion. This depth range and shot size were adequate for obtaining direct and reflected waves of the necessary amplitude. Therefore, all measurements of direct and reflected waves were made using SPM-16 geophones.

A resistive voltage divider was used at the outputs of the amplifiers to decrease the amplitude of the direct signal [4].

The geophones used in recording the direct waves were also used in recording reflected waves; in this case, the amplifier output was not attenuated.

An SPM-16 geophone clamped against the wall of a drill hole was used to detect the direct wave at interior points in the medium. The geophone case was sealed with epoxy resin to exclude moisture. In most cases, in observing direct and reflected waves to depths of 100 m, the hermetic seal was adequate.

A pneumatic bridging system was used to clamp a three-component geophone against the wall of the well bore with a consistent pressure. A cross section of the pneumatic bridge is shown in Fig. 1.

The basis for operation of the equipment is as follows: the assembly is lowered into the well to the desired depth, and the resin chamber P is pressured with air. The chamber expands, pressing the case with the geophone coils against the mass of the well bore. The pressure against the wall is controlled by the air pressure. The system is released by releasing the air from the resin chamber.

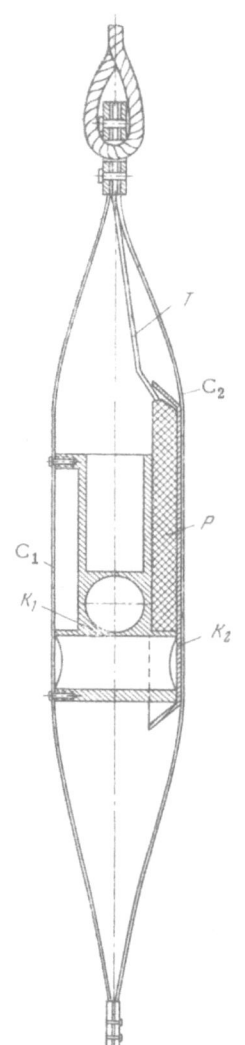

Fig. 1. Cross section of a three-component bridging well geophone. k_1) Frame for the emplacement of geophone coils in three orthogonal directions; k_2) support for resin chamber P; T) tube for supplying air to the resin chamber P.

This system permits measurements to be made at several depths in a well without removing the assembly.

The depth to which the assembly may be used is limited by the pressure limit of the plastic tubing. In our downhole studies of direct waves, the greatest depths were 50 m (the plastic tubing collapsed at 7-atm pressure).

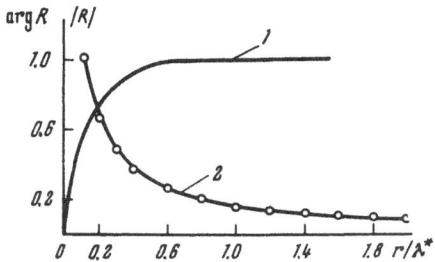

Fig. 2. Ratios of amplitude (1) and phase (2) to various r/λ* for spectral functions (1a) and (1).

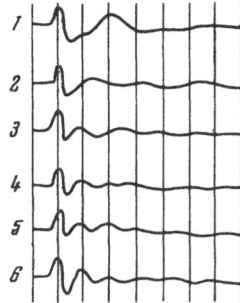

Fig. 3. Group of records of the direct longitudinal wave recorded with X-geophones at a depth of 10 m, with the detonation of a single squib. 1) x = 1.4 m; 2) x = 3.1 m; 3) x = 4.4 m; 4) x = 5.9 m; 5) x = 7.2 m; 6) x = 12.7 m; h_{shot} = 10 m.

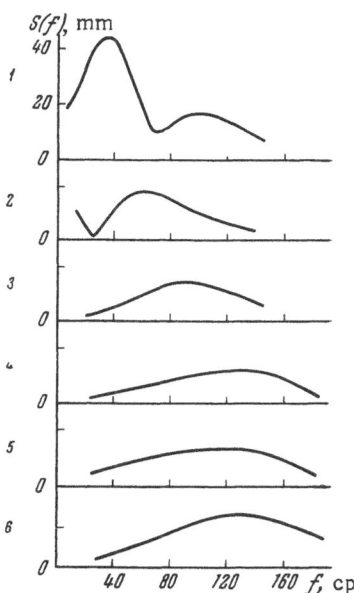

Fig. 4. The spectrums for the direct waves shown in Fig. 3.

The well geophone was oriented in the horizontal plane with the loading poles used in placing. However, this method could be used only at depths less than 20-25 m. At greater depths, no attempt was made at orientation.

In order to obtain components from the amplitudes of the recorded waves, it is necessary to use specially selected and matched geophone coils in the instrument. The geophones are selected by testing them with single electrical impulses, observing the natural oscillations of the system on an oscilloscope. The vertical and horizontal geophones are considered identical when the amplitude spectrums of the impulse response differ by no more than 1.5-2% while the phase shift has to be within limits which provide accurate timing on the records.

Recording Direct Longitudinal Waves at the Surface and Within a Medium

At the beginning of this paper, we enumerated the specific requirements for recording direct waves. Below, we will describe the effect of each of these factors on the form and spectrum of the direct wave. The study was carried out for cases of recording direct waves both at the surface and at interior points in the medium.

Variation of the Wave Form of the Direct Wave with Distance from the Source.
The direct waves have a number of specific properties at distances of less than a wavelength from the shot point. It has been pointed out in reference [3] that the form of the direct wave at short distances from the shot point changes with distance.

The important causes for the change in wave form with distance are as follows: 1) the process of wave-form development close to the source, where the effects of various terms in the propagation expression are felt [10]; 2) the effect of multipath arrivals complicating the direct wave [13]; 3) variation of shooting conditions with depth (in the case of the use of a fixed geophone location and varying shot depths) [3]; and 4) the attenuation of waves on travel to the geophone from the source [4] All of the enumerated factors which affect the form of the direct wave have been considered in the references cited. Here, we will consider the question of the transient spectrum of the direct wave, as it is affected by the various terms in the propagation function. It has been shown [10, 12] what the wave form is for a point source in an infinite, nonattenuating medium at a distance λ (λ is the apparent wavelength). However, it does not follow that at this distance, the complex spectrum of the specified pulse is also known over the entire frequency spectrum.

The displacement field developed by a point source in a homogeneous space may be written in the form

$$U_P = -\frac{1}{4\pi V_P^2 r}\left[\frac{1}{r}\Phi\left(t-\frac{r}{V_P}\right)+\frac{1}{V_P}\Phi'\left(t-\frac{r}{V_P}\right)\right],\tag{1}$$

where V_P is the longitudinal wave speed, r is the distance from the source to the measurement point and $\Phi(t-r/V_P)$ is the displacement potential.

At a fixed frequency, the first term in Eq. (1) quickly diminishes with increasing distance and beginning with some distance r, it may be ignored, and we can use the asymptotic expression

$$U_P \approx -\frac{\Phi'\left(t-\frac{r}{V_P}\right)}{4\pi V_P^3 r}.\tag{1'}$$

In determining this distance, we form the ratio of the displacement spectrum without consideration of the first term to the exact displacement spectrum:

$$R(\omega, r) = \frac{S_1(\omega)}{S_2(\omega, r)} = \frac{\frac{1}{V_P}j\omega S(\omega)}{\frac{1}{r}S(\omega)+\frac{1}{V_P}j\omega S(\omega)},\tag{2}$$

where $S_1(\omega)$ is the complex spectrum of the signal without consideration of the first term in Eq. (1), $S_2(\omega, r)$ is the exact expression for the complex spectrum, and $S(\omega)$ is the complex spectrum for the function $\Phi(t-r/V_P)$.

The modulus and argument of Eq. (2) are

$$|R(\omega, r)| = \frac{|r/\lambda^*|}{\sqrt{\left(\frac{r}{\lambda^*}\right)^2+\frac{1}{4\pi^2}}},\tag{3}$$

$$\arg R(\omega, r) = \Delta\varphi = \arctan\frac{\lambda^*}{2\pi r},\tag{3'}$$

where $\lambda^* = V_P/f$ is the circular wavelength.

Curves for the dependence of $|R|$ on $\Delta\varphi$ and r/λ^*, constructed using Eqs. (3) and (3') are shown in Fig. 2. It may be seen from this illustration that curve 1 quickly approaches its asymptote with increasing r/λ^*. For an accuracy of ±5% in frequency analysis, it may be considered that starting with a spacing $r = 0.4\lambda^*$, the amplitude spectrum of the pulse is established for the frequency range $f > V_P/0.4\lambda^*$.

Considering curve 2 in Fig. 2, the phase spectrum of the pulse is established at distances $r \gg 0.4\lambda^*$ from the source. Hence, we have an important result that in the low-frequency range, there are always frequency and phase components for which the asymptotic expression (1') is not valid. With increasing distance from the source, this region diminishes.

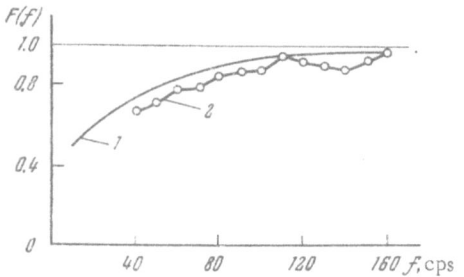

Fig 5. Curves for the ratios of spectrums for the direct wave at various distances from the source ($r_1 = 12.7$ m, $r_2 = 5.9$ m). 1) Theoretical curve; 2) observed curve.

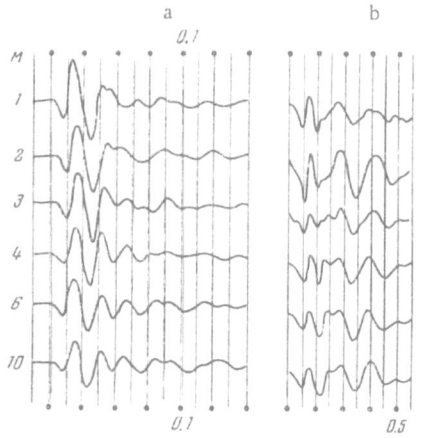

Fig. 6. Examples of records of the direct (a) and reflected (b) waves at various distances from the mouth of the shot hole. $h_{shot} = 20$ m; Q = 2 detonators.

The computed results were compared with experimental data on the change in the amplitude spectrum of the direct wave with distance from the source. The field method permitted recording of the direct waves without secondary pulses. The measurements were made along a horizontal profile, with X-geophones located beneath the low-speed layer. The geophone plant conditions were identical in all cases, and the instruments were held against the walls of the boreholes with uniform pressure. Vibrations were excited by the detonation of a squib at a fixed depth, equal to the depth of burial of the geophones.

A group of recordings of longitudinal direct waves recorded with X-geophones at a depth of 10 m with the detonation of a squib at 10-m depth is shown in Fig. 3, and the amplitude spectrums for these waves are shown in Fig. 4.

The direct waves at distances of 1.4 and 3.1 m from the shot point have low-frequency complications in the tail, which are caused by interference between the direct longitudinal wave and low-frequency plastic waves [13]. The corresponding spectrums also differ significantly from the spectrums of waves at larger distances from the shot point.

Beginning at a distance of 5.9 m from the source, the plastic waves are absent from the records, and the direct wave along with its spectrum changes only slightly.

We will not compare the theoretical and experimental curves for the ratios of spectrums for pulses at distances of 12.7 and 5.9 m (Fig. 5).

The theoretical curve was computed from the expression

$$|F(f)| = \frac{|S(f, r_1)|}{|S(f, r_2)|} = \sqrt{\frac{\left(\frac{2\pi f}{V_P}\right)^2 + \frac{1}{r_1^2}}{\left(\frac{2\pi f}{V_P}\right)^2 + \frac{1}{r_2^2}}}, \tag{4}$$

where $r_1 = 12.7$ m and $r_2 = 5.9$ m are the distances from the source to the observation points, and f is the frequency in cps.

It is apparent from a comparison of these curves that they agree quite well. It then follows that the assumptions made earlier about the transition to asymptotic conditions for components of the direct wave are confirmed by the experimental data.

In specifying the distance r, it may be assumed that this distance describes the surface of an equivalent sphere, the radius of which depends on the dimensions and size of the charge detonated [14]. However, this question requires special consideration.

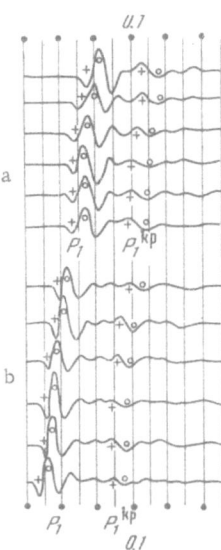

Fig. 7. A group of records of the direct wave (P_1) and a multiply reflected wave (P_1^{kp}) at the surface (a) and at the base of the low-speed layer (b). h_{shot} = 84 m; Q = 0.4 kg.

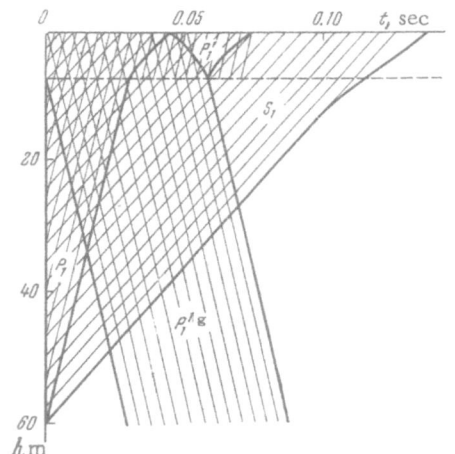

Fig. 8. Schematic ray paths for waves in the early part of a record.

Choice of the Optimum Spacing of the Geophones from the Mouth of the Shot Point. In recording the direct waves at the surface close to the mouth of the shot hole, it has already been noted that the form of the direct wave changes with minor changes in offset distances [3]. The reason for this is not apparent.

A special set of experiments was carried out to test the hypothesis that this behavior, the variation in form of the direct wave, is related to a disturbance of the medium as a result of drilling. In such a case, similar changes would have to be noted with the reflected waves.

Direct and reflected waves were recorded of distances of 1, 2, 3, 4, 6, and 10 m from the mouth of the shot hole. The geophones were buried at the same depths (20 cm) in packed wet sand. Similarity of geophone response was tested by recording reflected waves from a shot point 200 m from the geophone line. Geophone plants were considered to be identical when the phase shifts between them were no more than 0.001-0.002 sec, and the forms of the reflected waves appeared good on records from the various geophones.

When the direct waves were recorded, the geophones were not moved, but the charge was fired in a drill hole 1 m from the nearest geophone. Records of the direct wave at various distances from the mouth of the shot hole are shown in Fig. 6. A change in the form of the pulse with distance is seen. The change is especially apparent in the tail of the wave. The form of the wave changes insignificantly over distances of 4-10 m. These results are similar to those of reference [3].

The reflected waves were recorded along with the direct waves (see Fig. 6). A comparison of the various reflected waves shows that they, as well as the direct waves, change in form over distances of 1-3 m. The reflected waves are practically invariant over distances of 4-10 m.

Consequently, the change in form of the direct wave close to the mouth of the shot hole is related to disturbance of the medium as a result of drilling.

On Distortion of the Direct Wave as a Result of Interference with Waves of Other Types. The following wave types may be recorded in the early part of the record at points on the surface or within the earth in addition to the direct wave: 1) Multiply reflected waves which undergo one or several reflections between the base of the low-speed layer and the surface; 2) surface waves; and 3) direct shear waves.

Under certain conditions (of shot depth, wave speeds, layer thicknesses, and so on), these waves may interfere with the direct longitudinal wave, distorting part of the record. We will

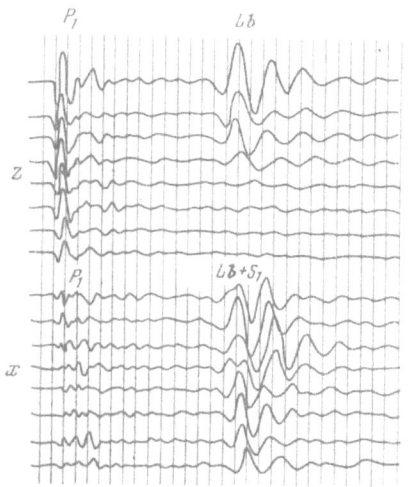

Fig. 9. Group of records of the direct wave (P_1), a surface wave (Lb), and a shear wave (S_1) on the Z and X components. $h_{shot} = 53$ m; $Q = 1$ detonator.

consider the properties of these waves and the conditions under which they may interfere with the direct wave.

Multiply Reflected Waves. The P_1 wave, which has a relatively simple form, on transmission through the low-speed layer is converted to a complicated P_1' wave, consisting of the P_1 wave itself and a multiply reflected wave P_1^{kp} (see Fig. 7). The diagnostic properties of the P_1^{kp} wave are as follows:

1. The P_1^{kp} wave is recorded for all shot depths;

2. the P_1^{kp} wave is reversed in phase in comparison with the P_1 wave; and

3. with observations at the base of the low-speed layer, there is a wave similar to the P_1^{kp} wave, but reflected from the Earth's surface (Fig. 7b). The differences in arrival times for the P_1 and P_1^{kp} waves are practically the same at the surface and at the base of the low-speed layer.

The direct wave at the surface or at the base of the low-speed layer may be distorted by interference with the multiply reflected wave when the two-way travel time for the wave in the low-speed layer is less than the duration of the corresponding direct wave P_1.

With measurements beneath the low-speed layer, the direct wave is followed immediately by a wave P_1^g (from Gamburtsev, Kuznetsov, and Isaev, next article). This is a compound vibration, consisting of the superposition of waves reflected from the top and bottom of the low-speed layer. The direct wave is confused with the P_1^g wave when the duration of the P_1 wave exceeds the difference in arrival times for the wave reflected from the base of the low-speed layer and the direct wave.

Figure 8 shows schematically the time ranges for ray paths of the principal waves recorded in the initial part of a record, for locations of the source and geophones in the depth range 0-60 m (with the source below the geophones). This chart was prepared on the basis of the structure for the medium described in the paper by Gamburtsev and Koptev in this volume. With this chart, and being given the duration of the direct wave, we can determine the conditions under which there is no interference between the direct wave and other types of waves.

Surface and Shear Waves. In recording the direct wave at a free surface at small distances from the mouth of the shot hole, a low-frequency surface wave Lb is recorded (Fig. 9a). The main properties of these waves are as follows:

1. The wave Lb is clearly traceable on the record from the vertical geophones; its arrival time is close to the arrival time for the direct shear wave S_1 recorded on the horizontal geophones;

2. a rapid decay with distance is observed for the Lb wave;

3. the intensity of the Lb wave and the area over which it may be traced are increased with increased shot depth; and

4. dispersion in propagation speeds is observed.

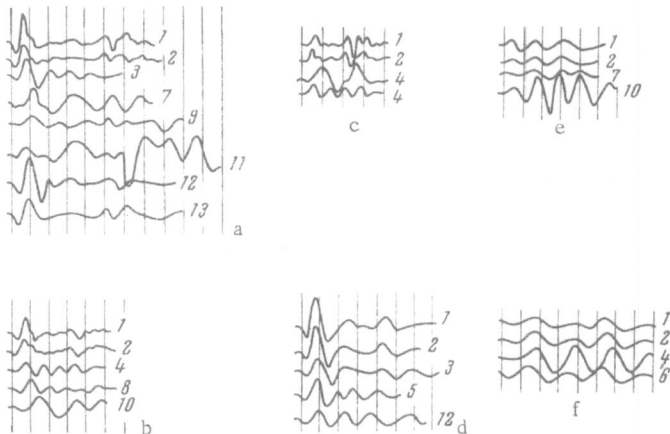

Fig. 10. a) SF 0, h = 30 m, Q = 1 detonator; b) SP-0, h = 23 m, Q = 1 detonator; c) SP-0, h = 19 m, Q = 1 detonator; d) SP-39, h = 32 m, Q = 0.075 kg; e) SP-0, h = 0, vibrator; f) SP-15, h = 14 m, Q = 1.2 kg; 1) the geophone was packed in a drill hole with wet sand; 2) the geophone was packed in a drill hole with tamped wet sand; 3) the geophone was on the axis of a water-filled hole, suspended by a cable; 4) the same, but in a dry hole; 5) the geophone was on the axis of a water-filled hole, floating in oil; 6) the same, in a dry hole; 7) the geophone was hung in a water-filled hole on a cable, in contact with the wall; 8) the same, in a dry hole; 9) the geophone was floating in oil in a water-filled hole, in contact with the wall; 10) the same, in a dry hole; 11) the geophone was suspended in a water-filled hole on a cable, not touching the wall; 12) the geophone was pressed against the wall of the water-filled bore (0.6 atm); 13) the same (1.6 atm).

It is apparent that the surface wave is a Lamb wave at the free surface [6]. With shots at very shallow depths, these waves interfere with the tail of the direct wave.

The shear wave S_1 at the free surface is recognized on records of the X-component. The S_1 wave is superimposed on the Lb wave at short distances from the mouth of the drill hole and for the appropriate shot depths (see Fig. 9b).

At interior points of the medium, the S_1 wave is recorded on both the horizontal and the vertical geophone. It is possible to determine the interference zone between the P_1 and S_1 waves inside the medium using the chart shown in Fig. 8.

<u>The Effect of Geophone Siting Conditions in a Drill Hole on the Recorded Form of Direct Waves.</u> The evaluation of the form of the direct wave recorded at internal points of a medium requires the selection of the best possible method for placing geophones in drill holes. A series of special experiments were carried out to study the effects of planting methods for geophones in drill holes. Measurements were made with geophones placed in drill holes by various means. The optimum conditions for siting geophones were taken to be those for which the clearest and most resolvable records were obtained.

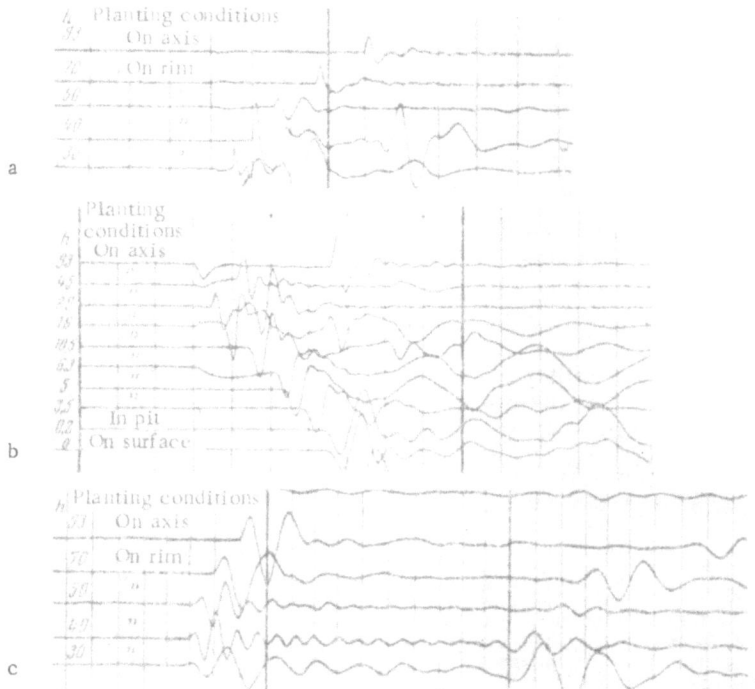

Fig. 11. Examples of records of the direct wave with
geophones located in drill holes with the arrival at the
geophones of high-frequency (a, b) and low-frequency
(c) energy. a, b) Q =1 detonator, h = 25 m, Δx = 3 m; c)
Q = 0.3 kg, h = 25 m, Δx = 50 m.

Experiment 1. The geophones were placed in shallow holes (7-8 m), located on an arc
of a circle at a radius of 5 m. The geophone plant conditions varied (see below), but the depth
of burial was nearly constant. Shot points were located at the center of the circle (SP-0) and
at distances of 15 m (SP-15) and 39 m (SP-39) from it. Vibrations were excited by detonation
of a blasting cap or explosive charges of various weights, as well as with a vibrator (the vibrator was placed on the ground about the mouth of the hole located at the center of the circle,
V-0).

Experiment 2. The geophones were located in a drill hole spaced at intervals of 2-3 m,
suspended in drilling mud. The depth of burial ranged from 3-93 m. The geophones in the
shallow part of the hole were implanted with wet sand. The geophones in the deeper part of
the hole were packed in place with drilling mud. Vibrations were excited using electrical detonators or explosive charges in holes at a distance of 3 m (SP-3) or 50 m (SP-50).

Results: Records of the direct wave obtained in experiment 1 are shown in Fig. 10 (details of the experimental setup are given beneath the illustration). The records obtained in experiment 2 are shown in Fig. 11. It is apparent from these illustrations that the effect of plant
conditions on the form of the recorded direct wave differs significantly for high-frequency and
low-frequency waves. We will consider these two cases.

1. The highest-frequency signals were generated by the detonation of an electrical detonator beneath the base of the low-speed layer. The maximum in the direct wave spectrum
was located at frequencies of 150-200 cps in this case. The sharpest and most resolvable
records for the direct longitudinal and shear waves were obtained with the following plant

conditions: a) the geophones were located on the axes of shallow holes, packed in dry or wet sand (Fig. 10: 1, 2); b) the geophones were pressed tightly to the wall of the bore (Fig. 10: 12, 13): and c) the geophones were located on the axis of the hole, in a relatively deep hole filled with drilling mud (Fig. 11a, b).

For the other planting conditions used in experiment 1, the records of the direct wave were poorer (the P_1 and S_1 waves are of lower frequency and are recorded with considerable noise). The records from geophones suspended in drilling mud beneath the low-speed layer (experiment 2) are complicated to a lesser degree than are the records from the geophones suspended in water within the low-speed layer, and it is possible to positively identify the direct shear and longitudinal waves on them.

2. Low-frequency signals (with a maximum in the spectrum at 80 cps) were recorded on geophones at interior points of the medium for detonation of charges of 0.3-2.5 kg, and for the vibrator. It is apparent from Fig. 10d, e, and f and Fig. 11c that with the arrival of low-frequency energy at a geophone, the effect of the geophone plant is still significant but not nearly as large as in the case of high-frequency energy.

Thus, the sharpest and most resolvable records of the direct wave were obtained with placement of the geophones on the well axis or pressed against the wall of the bore. As indicated by this work, in describing the placement of geophones in holes, the fact that records of the direct wave are good is evidence for identical planting conditions for the geophones.

Conclusions

We have dealt with questions concerning the recording of direct longitudinal waves on the surface and at interior points of a medium.

1. In determining the spectrum of a direct wave, such that asymptotic expressions are valid over a wide frequency range, the observation points must be located at distances $r \geq 0.4\lambda^*$ (where λ^* is a circular wavelength).

2. In recording the direct wave at the surface, it is necessary to make measurements at distances of more than 2-4 m from the mouth of the well. In this case, the direct wave is not distorted by the effects of drilling.

3. In recording the direct wave either on the surface or at interior points, the measurement point must be outside the interference zone with other waves. It is necessary to know the duration of the direct wave and all propagation speeds for waves recorded in the early part of the record in order to determine the interference zone.

4. Geophones at interior points of the medium must either be placed on the well axis (in dry, shallow wells the geophones must be packed in dry or wet sand, and in deeper holes, the geophones must be packed in mud) or they must be pressed tightly against the wall of the bore hole.

5. Two-level recording may be used to record the direct wave and reflected waves simultaneously. In this case, the effects of geophone siting can be avoided by a comparative analysis of the direct and reflected waves.

The experiments for determining the filtering properties of the low-speed layer described in another paper in this collection may serve as an example of the use of methods for recording direct waves.

It should be noted that methods for recording direct longitudinal waves at the surface or in holes cannot be considered to be fully worked out. In particular, questions dealing with the effect of the various terms in the propagation equation on the form of the direct wave and the effect of plastic waves have not been considered.

LITERATURE CITED

1. E. I. Gal'perin, Description of a detailed study of velocity models for the upper part of the section under conditions of weak velocity differentiation, Izv. Akad. Nauk SSSR, Ser. Geofiz., No. 4 (1964).

2. N. I. Berdinnikova, On some effects of anisotropy in layered media with work on shear waves, in: Aspects of the Dynamic Theory of Propagation of Seismic Waves, No. 2, Leningrad State University, 1959.

3. A. M. Epinat'eva, V. V. Kuznetsov, Yu. A. Ostrovskii, and L. L. Khudzinskii, Some experimental data on the form of pulses generated by shots in drill holes, Izv. Akad. Nauk SSSR, Ser. Geofiz., No. 6 (1963).

4. V. V. Kuznetsov, Determining the average coefficients for attenuation and reflection coefficients from the spectrums and amplitudes of direct and reflected waves, Tr. Inst. Fiz. Zemli, Akad. Nauk SSSR, No. 34:201 (1964).

5. A. V. Nikolaev, Properties of elastic wave fields close to a vibrator source on a free surface, Izv. Akad. Nauk SSSR, Fiz. Zemli, No. 3 (1965).

6. I. S. Berzon, Some results of the study of seismic waves from well shooting, Izv. Akad. Nauk SSSR, Ser. Geofiz., No. 9 (1964).

7. V. V. Zhadin, Investigation of the decay and dispersion of seismic waves using well surveys, in: Aspects of the Dynamic Theory of Propagation of Seismic Waves, No. 2, Leningrad State University, 1959.

8. N. I. Berdennikova, V. V. Zhadin, and A. G. Rudakov, To the question of the methods of seismic well surveys, in: Aspects of the Dynamic Theory of Propagation of Seismic Waves, No. 2, Leningrad State University, 1959.

9. E. I. Gal'perin and A. V. Frolova, Three-component seismic survey in a well, Izv. Akad. Nauk SSSR, Ser. Geofiz., No. 6 (1961).

10. I. P. Kosminskaya, Amplitude curves and phase travel times for seismic waves generated by a point source and propagating in a homogeneous ideally elastic infinite space, Izv. Akad. Nauk SSSR, Ser. Geofiz., No. 6 (1952).

11. L. L. Khudzinskii, Broadband seismic equipment ShPSS, Izv. Akad. Nauk SSSR, Ser. Geofiz., No. 4 (1963).

12. I. I. Gurvich, Seismic Exploration, Gostoptekhizdat, 1960.

13. Yu. I Vasil'ev and M. N. Shcherbo, Plastic wave motion in the ground, Izv. Akad. Nauk SSSR, Ser. Fiz. Zemli, No. 10 (1965).

14. I. I. Gurvich, To the theory of spherical radiation of seismic waves, Izv. Akad. Nauk SSSR, Fiz. Zemli, No. 10 (1965).

THE POSSIBILITY OF DETERMINING
THE FILTERING PROPERTIES
OF THE UPPER PART OF THE SECTION

A. G. Gamburtsev, V. V. Kuznetsov, and V. S. Isaev

The low-speed layer in a real medium is highly inhomogeneous with respect to wave speeds and has an important effect on the character of the wave field [1-3]. The zone of low wave speeds may be considered to be a single layer or a series of thin layers. A study of the effect of this zone on waves of various types is important for the following reasons.

1. Usually seismic exploration is carried out with geophones located on the Earth's surface; all deeper waves have to pass through the low-speed layer and are filtered to a greater or lesser degree by it.

2. Frequently multiple reflections and refracted waves which undergo one or several intermediate reflections in the low-speed layer are included on seismic records; moreover, waves reflected from the low-speed layer may be recorded by geophones located within the medium.

3. Because of the comparative accessibility of the upper part of the section and in particular, the low-velocity zone, it is possible to examine this medium as a distinct unit, the study of which is necessary to aid in the study of the deeper parts of the section.

The filtering properties of the low-speed layer, as with any thin layer, may be represented in the form of a frequency response function [4].

The present paper is a development of a method for finding the response function for the low-speed layer for the transmission and reflection of longitudinal waves for normal incidence. The results of such determinations and comparison with theory are described. Methods for and the results of the determination of the relationship between attenuation factor and frequency in the low-speed layer and in the underlying beds are described.

In the field surveys, a great many measurements were made on the surface and at interior points of the medium. A description of the field procedures used in the studies has been given in the paper by Kuznetsov and Gamburtsev, here presented on p. 163. We will deal only briefly with the basic features of the field methods.

1. Direct and reflected waves were recorded on the surface and at interior points of the medium.

2. Geophones were placed in drill holes on the hole axes. For measurements above the fluid level in the holes (within the low-speed zone), the geophones were packed in wet sand. For measurements in deeper holes (up to 100 m), the geophones were hung in drilling mud.

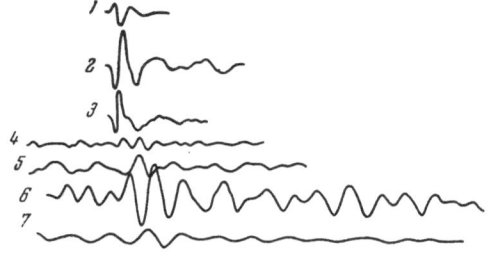

Fig. 1. Examples of waves recorded at the surface and at a depth of 100 m. 1) Wave P_1, h = 100 m; 2) wave P'_1, h = 0 m; 3) wave P'^g_1, h = 100 m; 4) reflected wave $P_{0.3}$, h = 100 m; 5) reflected wave $P_{0.3}$, h = 0 m; 6) reflected wave $P_{1.3}$, h = 0 m; 7) reflected wave $P_{1.3}$, h = 100 m.

3. In the surveys, waves were recorded with several geophones located at the same depth, so that data could be stacked.

4. The geophones and shot holes were laid out to ensure practically vertical incidence in the low-speed layer.

5. The distance from the shot holes to the geophones and the charge sizes were selected so that secondary pulses were not present on the direct-wave record and so that the records were not affected by the proximity of the shot hole (that is, so that asymptotic expressions would be valid [5]).

The recording was done with the ShPSS-48 model seismic equipment. SPM-16 geophones were used.

Data on the wave-speed profile in the near surface rocks have been given by Gamburtsev and Koptev in the next paper.

Effect of the Low-Speed Layer on the Records of Direct and Reflected Waves

Significant differences were observed in the wave forms, amplitudes, and frequencies of the seismic waves recorded at the surface and at various depths in drill holes. These differences were particularly apparent on recordings of the direct waves. The direct waves P_1 which were recorded at some depth beneath the low-speed zone consisted of simple vibrations lasting for 1.5-2 cycles. The maximum in the spectrum was between 130 and 200 cps.

Extensive interference vibrations could be traced on the early parts of the records made at the surface and within the low-speed zone. The first of these vibrations to be recorded was the purely direct wave P_1. This wave was similar in appearance and general form to the P_1 wave recorded beneath the low-speed layer, though the latter was somewhat higher in frequency. Immediately after the P_1 wave, a combination of vibrations caused by multiple reflections between two strongly reflecting boundaries, the top and bottom of the low-speed layer, can be identified. Some role is played also by multiple reflections on wave-speed discontinuities within the low-speed zone.

Thus, the low-speed zone transforms a simple vibration P_1 into a complex vibration, which will be designated as the P'_1 wave. The low-speed zone has a similar effect on reflected waves formed at greater depths and transmitted back through the low-speed layer.

A deep reflection also appears as a complex combination of vibrations, P'^g_1 (Fig. 1). The index "g" indicates a ghost arrival corresponding to the direct wave [10]. The first part of this vibration, the wave P^g_1, is a reflection from the base of the low-speed layer, as though from a thin layer (see below), and is similar in form to the P_1 wave recorded at the same point. The later part of the vibration is complicated, and as in the preceding case, may be considered to be formed by multiple reflections.

Figure 1 shows typical records of the P_1 and P'^g_1 waves for a geophone burial depth of 100 m, and the wave P'_1 at the surface.

TABLE 1

Recording location	Parameter, cps	Direct P_1	Reflections	
			$P_{0.3}$, $P_{0.4}$	$P_{1.3}$
At the surface	f apparent	70-80	60-70	50
	f max (from the spectrum)	65	55	45
	Absolute width of the spectrum at a level of 0.7 A_{max}	45	42	33
At the base of the low-speed zone	f apparent	100-110	80-100	60
	f max (from the spectrum)	92	70	50
	Absolute width of the spectrum at a level of 0.7 A_{max}	70	46	33

Thus, the low-speed zone acts as a thin layer which transforms a comparatively simple wave form P_1 into a complex vibration; P_1' at the surface and $P_1'g$ below the low-speed zone.

Generally, other types of waves arriving at the low-speed zone from below to be reflected or transmitted must undergo similar modifications to become more complicated vibrations. However, it is difficult to observe this behavior on seismic records for the following reasons.

1. The direct wave at the surface or in a drill hole, and the $P_1'g$ wave in a drill hole are recorded at a time of low noise, while the reflected waves are recorded during periods of high noise (Fig. 1);

2. the reflected wave traceable on the seismic records has a longer duration in time than the direct wave, and therefore the multiple reflected wave formed in the low-speed zone interferes with the reflected wave which travels through the zone only once; and

3. recognition of the "tail" of a reflected wave with geophones at the surface or within the low-speed zone is confused by ghost arrivals.

Therefore, we may conclude that the direct waves recorded at the surface (P_1') and beneath the low-speed zone (P_1) or the ghost ($P_1'g$) recorded beneath the low-speed layer will be the most satisfactory for studying filtering in the low-speed zone.

We will now consider the effect of the low-speed zone on waves with different spectral compositions. The highest-frequency waves (direct longitudinal waves and reflected waves with short transit times) are expressed most sharply and are best resolved on the records obtained within the medium, and especially below the low-speed zone. Differences in recordings of lower-frequency waves (later reflections) at the surface and in wells are less significant; the lower the apparent frequency of a wave, the less will be the difference in the character of its recording at interior points and at the surface.

Table 1 lists some data on apparent frequencies and other spectral characteristics for various waves.

It follows from this that the low-speed zone has a significant effect on the form, duration and frequency content of the waves recorded at the Earth's surface and on reflections from the low-speed zone.

Theoretical Response Function for the Low-Speed Layer

The low-speed zone can be thought of as a special case of a thin layer in which the upper boundary is a free surface. The reflection factors at the top and bottom of the low-speed layer are much larger than those that we normally find deeper in the section. In view of this, it

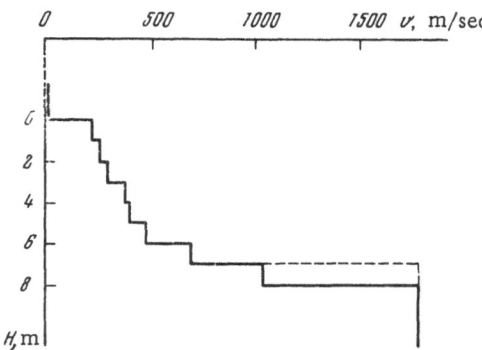

Fig. 2. Models of the medium selected for computations of the response function for reflection from the low-speed zone.

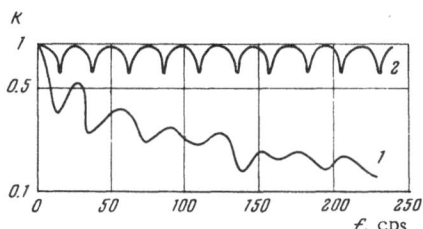

Fig. 3. Response functions for reflection from the low-speed zone. 1) Experimental data; 2) computed.

would be desirable to review briefly the theory for the frequency response for reflection from and transmission through the low-speed layer. In the following section, we describe some experimental results of the determination of the response function and present an evaluation of the attenuation coefficient in the low-speed layer and in the underlying rocks. We have limited our considerations to the amplitude response function.

On the Form of the Response Function. The response function for reflection from the low-speed layer may be written as [4]

$$k(\omega) = \frac{S_{\text{out}}(\omega)}{S_{\text{in}}(\omega)} = \eta(\omega)\psi(\omega), \tag{1}$$

where ω is the radian frequency, $k(\omega)$ is the respone function for reflection from the low-speed layer, $S_{\text{out}}(\omega)$ and $S_{\text{in}}(\omega)$ are respectively the wave spectrums at the output and the input to the layer, and $\eta(\omega)$ and $\psi(\omega)$ are characteristics respectively representing the inhomogeneity and the attenuation in the low-speed layer.

The response function for transmission of waves through the low-speed layer may be written as

$$d(\omega) = \frac{S'_{\text{out}}(\omega)}{S_{\text{in}}(\omega)} = \eta'(\omega)\psi'(\omega)\frac{Q_1(\omega)}{Q_2(\omega)}, \tag{2}$$

where $d(\omega)$ is the response function for transmission of waves through the low-speed layer, and $Q_1(\omega)$ and $Q_2(\omega)$ are respectively the response functions for the system geophone−soil (at the surface) or the system geophone−surrounding medium (in a well).*

Let us examine what the behavior of the response functions would be in the absence of attenuation and for $Q_1(\omega) = Q_2(\omega)$.

In this case

$$k(\omega) = \eta(\omega) \tag{1'}$$

and

$$d(\omega) = \eta'(\omega). \tag{2'}$$

We will consider a simplified case. Let a thin homogeneous layer (layer 2) lie on a half-space (1) and be bounded by a free surface. Let $\rho_1 V_1 > \rho_2 V_2 > \rho_3 V_3 = 0$.

---

*Similar response functions were not used in Eq. (1) because experimentally, the waves into and out of the layer are measured with the same geophone.

Then, the complex spectrum for the waves reflected from the low-speed layer may be written as [8]

$$\overline{S}_{\text{out}}(\omega) = \overline{S}_{\text{in}}(\omega)\,[k_{12} + (1 - k_{12}^2)\,e^{-j\omega 2\tau}\,(1 - k_{12}e^{-j\omega 2\tau} + k_{12}^2 e^{-j\omega 4\tau} - \ldots)], \tag{3}$$

where τ is the travel time for waves between the top and bottom of the low-speed layer, $\overline{S}_{\text{in}}(\omega)$ is the complex spectrum of the incident waves, k_{12} is the reflection factor from the base of the low-speed layer, and $k_{23} = 1$ is the reflection factor at the free surface.

Transforming to trigonometric functions, we can write Eq. (3) in the form

$$\overline{S}_{\text{out}}(\omega) = \overline{S}_{\text{in}}(\omega)\,\frac{\cos 2\omega\tau + k_{12} - j\sin 2\omega\tau}{1 + k_{12}\cos 2\omega\tau - jk_{12}\sin 2\omega\tau}. \tag{4}$$

It follows from Eq. (4) that the amplitude spectrum at the output from the layers is exactly equal to the amplitude spectrum at the input

$$S_{\text{out}}(\omega) = S_{\text{in}}(\omega), \tag{5}$$

but the phase spectrums for these waves differ.

Thus, the amplitude of the response function of the low-speed layer for wave reflection is unity, and does not depend on frequency, while the phase spectrum does depend on frequency.

The corresponding expressions for waves passing through the low-speed layer and being recorded at its surface are

$$\overline{S}_{\text{out}}(\omega) = \overline{S}_{\text{in}}(\omega)\,(1 + k_{12})(1 - k_{12}e^{-j\omega 2\tau} + k_{12}^2 e^{-j\omega 4\tau} - \ldots), \tag{3'}$$

$$\overline{S}_{\text{out}}(\omega) = \overline{S}_{\text{in}}(\omega)\,\frac{1 + k_{12}}{1 - k_{12}\cos 2\omega\tau + jk_{12}\sin 2\omega\tau}, \tag{4'}$$

$$S_{\text{out}}(\omega) = S_{\text{in}}(\omega)\,\frac{1 + k_{12}}{\sqrt{1 + k_{12}^2 - 2k_{12}\cos 2\omega\tau}}. \tag{5'}$$

Thus, it is apparent that the response function for transmission through the low-speed layer depends on frequency.

The result about the reflection coefficient for the low-speed layer being independent of frequency is valid only when all of the multiply reflected waves are considered in the spectral analysis of the reflected waves. Our expression for the response function for reflection from the low-speed zone for the case of repeated transmission of the waves into the low-speed zone is

$$k(\omega) = \sqrt{1 + 2k_{12}(1 - k_{12}^2)\cos 2\omega\tau + (1 - k_{12}^2)^2},$$

that is, the reflection factor does depend on frequency.

Response functions for reflection from the low-speed zone were evaluated on the BESM-2 computer. Two very similar models, as shown in Fig. 2, were used, which closely correspond to an actual case*(Gamburtsev and Koptev, next paper). The computations of the response function were made for eight-fold travel times for the waves in the layer. Similar results were obtained for both models. The response functions are shown in Fig. 3.

---

*All densities were taken as being equal.

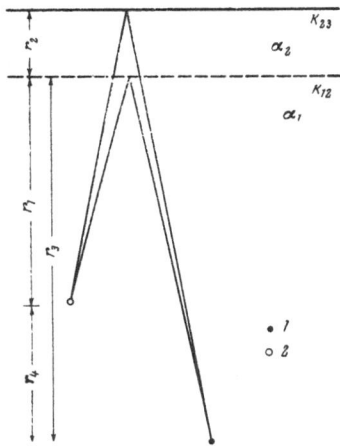

Fig. 4. Simplified ray paths taken for evaluating the effect of attenuation of the computation of response functions. 1) Geophones; 2) shot point.

<u>Fictitious Response Functions</u>. We will assume that a wave is recorded at the output of the layer which has travelled through the layer only once, not being subjected to multiple reflection in the layer (corresponding to the direct wave). Then, if there is no attenuation, the wave recorded at the output must be exactly the wave which was incident at the layer. In this case, the transmission factor does not depend on frequency. The same result may be obtained with reflection of the direct wave from the low-speed zone, if the wave reflected from the bottom of the zone can be separated from the rest of the record.

In relation to this, we may introduce the concept of a fictitious response function. It is determined by the ratio of the running spectrum [9] for the wave at the output of the layer to the spectrum of the wave incident at the layer. The fictitious response function has physical meaning and may characterize the properties of the layer with the following limitations:

1. There must not be any fully reflected waves propagating within the layer included in the pulse at the output of the layer which is selected for analysis; and

2. there can be no partially reflected waves propagating in the layer which affect the part of the record which is selected.

In the opposite case, these functions will be different for pulses of different lengths, and basically will be related to the properties of the signal rather than to the properties of the layer.

It is possible to determine the frequency dependence of the average attenuation factor in the low-speed zone from the fictitious response function for transmission through the low-speed zone, while the frequency dependence of the average attenuation factor in the underlying rocks may be determined from the fictitious response function for reflection from the low-speed zone. The expressions used in determining $\bar{\alpha}(f)$, as well as the results, will be given below.

Experimental Data on the Response Functions

from the Low-Speed Zone

The response function for transmission and for reflection by the low-speed zone will be expressed in the form of a ratio of the spectrum for the waves recorded at the output of the zone to the spectrum of the waves at the input to the zone.

We will assume that the waves at the input are waves arriving at the zone from depth, while the waves at the output are waves proceeding through the zone and recorded at the surface (transmission) and waves reflected from the zone (reflection).

With records of the direct wave at the base of the low-speed zone (the input), the trace is complicated by reflected waves arising within the zone. Therefore, it is preferable to use records made at great depths, rather than at the base of the zone, so that the direct wave will lie outside the zone of interference with multiples. In this case, the response function is distorted somewhat by inhomogeneities and distortion in the medium beneath the low-speed zone. In the area under consideration, the effect of these factors is not large, inasmuch as the underlying rocks are practically homogeneous and do not exhibit much attenuation (see below).

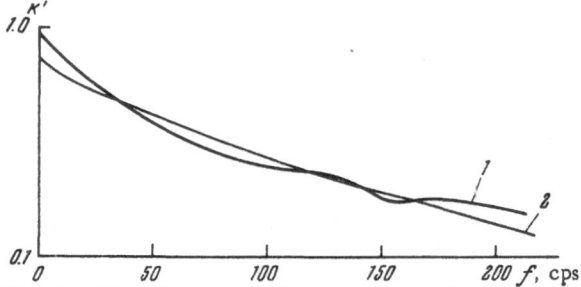

Fig. 5. Envelopes of the response function for reflection from the low-speed layer. 1) Experimental data; 2) computed, with attenuation considered.

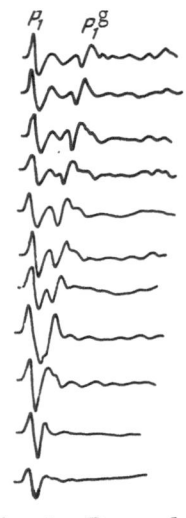

Fig. 6. Records of the P_1 and P_1^g waves as the shot point is moved toward the base of the low-speed zone. The shot depth changes from 27 to 7 m, while the geophone depth is 56 m.

A similar result may also be derived for the experimental response function for reflection from the low-speed zone. With a nearly homogeneous and nonattenuating region beneath the low-speed zone, waves should be recorded at a great enough depth beneath the zone that there is no interference with other waves.

The wave spectrums used in determining the response functions were obtained on a frequency analyzer [10] or on a BESM-2 computer.

Response Function for Reflection from the Low-Speed Zone. The response function for reflection from the low-speed zone was taken as the ratio of the spectrum of the waves reflected from the zone (P_1^g) to the spectrum of the incident waves (P_1), recorded on a single geophone located beneath the low-speed zone.

The wave spectrums at the input and output differ significantly. The form of the spectrum at the input is simple in most cases, and has a single dominant maximum. At the output, the spectrum is more complicated; several maximums and minimums are apparent.

Therefore, the positions of the extremals for the response function of the low-speed zone are determined primarily by the form of the wave spectrum at the output.

A response function for reflection from the low-speed zone, determined from experimental data, is shown in Fig. 3. A consideration of this curve leads to the following conclusions.

The experimental response function has an oscillatory form and decays with increasing frequency. The theoretical response function is also oscillatory but does not decay with frequency. The minimums on the computed functions are much sharper than the maximums. The frequencies at which experimental values may be recognized are quite close on both characteristics. From a comparison of the curves, it follows that inhomogeneity in the low-speed zone does not result in attenuation of the response function with frequency. Consequently, the decaying form of the experimental response function must be explained by the existence of attenuation in the low-speed zone or in the underlying rocks.

We have made a detailed comparison of the experimental data with theory, taking into consideration the effect of attenuation in the low-speed zone and in the underlying rocks. The calculations were carried out for a thin layer without consideration of multiples (Fig. 4).

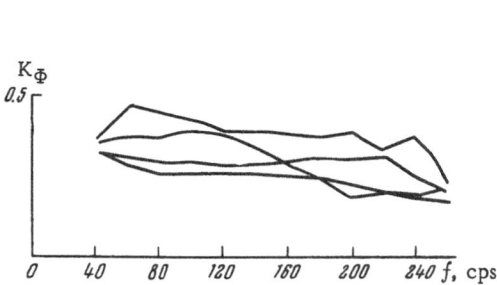

Fig. 7. Examples of some fictitious response functions for reflection from the low-speed zone.

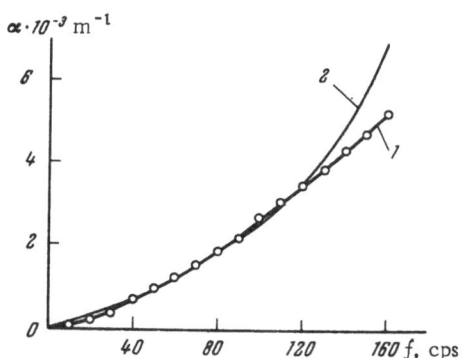

Fig. 8. Curve for the relationship $\bar{\alpha}_2 = \bar{\alpha}_2(f)$. 1) Experimental; 2) computed.

The theoretical response function, with attenuation being considered, was computed as

$$k(f) = \frac{S_{\text{out}}(f)}{S_{\text{in}}(f)}.$$

The value for $S_{\text{in}}(f)$ was represented as

$$S_{\text{in}}(f) = S_0(f)\frac{e^{-\bar{\alpha}_1(f)r_4}}{r_4}\,Q(f),$$

while the value for S_{out} was determined from the expression (see reference [9], for example)

$$S_{\text{out}}^2(f) = S_0^2(f)\,Q^2(f)\left[k_{12}^2\,\frac{e^{-2\bar{\alpha}_1(f)(r_1+r_3)}}{(r_1+r_3)^2} + \right.$$

$$\left. + 2k_{12}k_{23}(1-k_{12}^2)\,\frac{e^{-2\bar{\alpha}_1(f)(r_1+r_3)-2\bar{\alpha}_2(f)r_2}}{(r_1+r_3)(r_1+r_3+2r_2)}\,\cos 2\omega\tau + (1-k_{12}^2)^2\,k_{23}^2\,\frac{e^{-2\bar{\alpha}_1(f)(r_1+r_3)-4\bar{\alpha}_2(f)r_2}}{(r_1+r_3+2r_2)^2}\right],$$

where $\bar{\alpha}_1(f)$ and $\bar{\alpha}_2(f)$ are the average attenuation factors beneath the low-speed zone and in the zone, respectively.

The values for τ, k_{12}, and k_{23} were computed for the models of the medium shown in Fig. 2. The values for $\bar{\alpha}_1(f)$ and $\bar{\alpha}_2(f)$ were determined experimentally (see below).

In this case, the response function is a decaying oscillatory curve. The envelopes of the theoretical and experimental curves were compared for correspondence about the maximums of the oscillatory curves. The experimental curve agrees well with the theoretical curve computed with consideration of attenuation (Fig. 5).

Fictitious Response Function for Reflection from the Low-Speed Layer. As noted earlier, the wave reflected from the low-speed layer $(P_1^{\prime g})$ is complicated in form. A relatively intense, high-frequency wave (P_1^g) may be traced at the beginning of this vibration. We will show that this wave is the reflection from the base of the low-speed zone.

1. It was mentioned earlier that the P_1^g wave is similar in form to the direct wave P_1 recorded at the same point (if the phase reversal is ignored). The spectrums of these waves are similar.

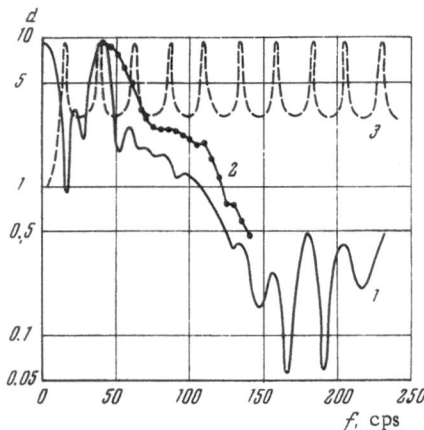

Fig. 9. Response function for wave transmission by the low-speed zone. 1) Experimental data obtained from spectrums of direct waves; 2) experimental data obtained using spectrums of the reflected wave $P_{0.3}$; 3) computed.

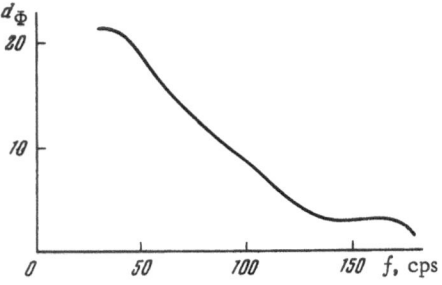

Fig. 10. Fictitious response function for wave transmission by the low-speed zone.

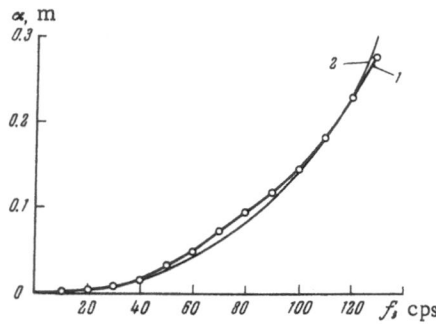

Fig. 11. Relationship between the average attenuation factor and frequency in the low-speed zone. 1) Experimental; 2) computed.

2. With the shot point close to the base of the low-speed layer, the direct wave and its ghost are close in time, and form a single pulse (Fig. 6).

Thus, it may be considered that the wave P_1^g is a reflection from the base of the low-speed zone as though from the boundary of a thin layer.

The fictitious response function was determined over the frequency range 40-250 cps (Fig. 7). The curve is smooth. The slope of the response function is apparently explained by attenuation in the medium beneath the low-speed zone. Thus, there is a possibility of determining $\overline{\alpha}_1 = \alpha_1(f)$ in this region.

Evaluation of the Frequency Dependence of Attenuation Beneath the Low-Speed Zone. A method described in reference [11] was used in determining the spectral behavior of the attenuation factor. Calculations were done using the expression

$$\overline{\alpha}_1(f) = \frac{1}{r_1}\left[-\ln k_\Phi(f) + \ln k_{12} - \ln \frac{r_3 + r_1}{r_3 - r_1}\right],$$

where $k_\Phi(f)$ is the fictitious response function for reflection from the low-speed zone.

Records of the direct wave and its ghost recorded at a depth of 100 m, with detonation of a charge of 0.4 kg at a depth of 70 m were used in determining the relationship of the attenuation factor to frequency. The direct wave and ghost were resolvable in time on these records.

The value for $\overline{\alpha}_1$ was determined over the frequency range 10-160 cps. At frequencies greater than 160 cps, the relationship between the attenuation factor and frequency was not determinable because the spectral components for the direct wave and ghost were too small.

The behavior of $\overline{\alpha}_1$ as a function of frequency in the rocks lying beneath the low-speed zone to a depth of 100 m is shown in Fig. 8. The experimental curve is the average of three sets of data.

We found the following expression which is an approximate representation of the relationship $\overline{\alpha}_1 = \overline{\alpha}_1(f)$:

$$\overline{\alpha}_1 = 0.29 \cdot 10^{-5} f^{1.43}\,\text{m}^{-1}$$

With this method, the reflection factor at the base of the low-speed zone may be determined [11],

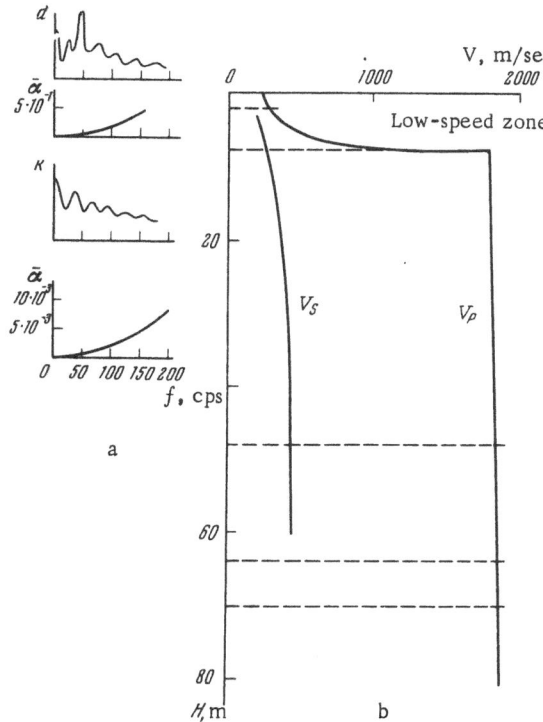

and in this case, it is 0.6. Thus, the base of the low-speed zone is a strongly reflecting interface.

Spectral Response for Transmission by the Low-Speed Zone. We used the method described below in determining the response function for transmission of waves by the low-speed zone.

The locations of the extremals in this function are controlled by the wave spectrum at the output of the layer.

Because the form of the wave spectrum at the output depends on the geophone planting conditions, it is preferable to use wave forms averaged from several geophones. In this case, we used wave spectrums recorded with several geophones located at the same distance from the mouth of the shot hole, and then averaged the spectrums.

Figure 9 shows the response function for transmission by the low-speed zone, determined from spectrums of direct waves recorded at the surface (the output of the layer) and at some depth beneath the base of the low-speed zone. Computed functions, based on the use of an expression given in reference [4]:

Fig. 12. Model of the medium. a) (From the top down) response function for transmission by the low-speed zone, frequency dependence of attenuation in the low-speed zone, response function for reflection from the low-speed zone, frequency dependence of attenuation in the underlying rocks; b) velocity profiles.

$$d(f) = \sqrt{q_{13}[1 - k^2(f)]},$$

where q_{13} is the ratio of acoustic impedances in the beds beneath the low-speed zone and above it, are also shown here.

It may be seen from these curves that, just as in the case of the response function for reflection from the low-speed zone, the observed behavior differs from theory in that the oscillatory curve decays with frequency.

The response function computed from the spectrum for a reflected wave $P_{0.3}$ recorded at a depth of 100 m and on the surface is shown on this figure for comparison. This wave is traceable on the records when the noise level is low. The frequency analysis of the signal recorded at the surface included consideration of the tail part of the vibration. Generally, this response function follows the function determined from the direct-wave spectrum, but the extremals are either not as sharp or are missing. The experimental response function for transmission by the zone is much steeper than the response function for reflection from the zone.

A comparison of the experimental curves with theoretical curves computed with attenuation being considered was made for the envelope of the oscillatory curves, in a manner similar to that described earlier. The envelopes in this case are similar in form over the frequency range 30-200 cps.

Fictitious Response Function for Wave Transmission by the Low-Speed Zone and Evaluation of the Attenuation Factor in the Low-Speed Zone. An example of a fictitious function determined from the spectrums of stacked records obtained at the surface and at the base of the low-speed zone is shown in Fig. 10.

The relationship $\bar{\alpha}_2 = \bar{\alpha}_2(f)$ was determined over the frequency range 30-130 cps for this function. Calculations were done using the expression

$$\bar{\alpha}_2(f) = \frac{1}{r_2} \left(\ln \frac{r_1}{r_1 + r_2} - \ln d_f \right),$$

where d_f is the fictitious response function for transmission by the low-speed zone. The relationship between the attenuation factor and frequency is shown in Fig. 11. Values for $\bar{\alpha}_2$ were obtained by extrapolation over the range of frequencies 0-30 cps. An empirical expression for the dependence of attenuation factor on frequency is

$$\bar{\alpha}_2 = 0.228 \cdot 10^{-5} f^{2.40} \, \text{m}^{-1}.$$

We might mention that in analyzing the spectrums for direct waves recorded along a vertical profile within the low-speed layer, nearly the same relationship $\bar{\alpha}_2(f)$ was found.

Comparison of the results of evaluation of the attenuation factor shows that it is larger by an order of magnitude in the low-speed zone than in the underlying rocks.

Models of the Near Surface

In constructing a representation for the near surface part of the section, it is necessary to have values for its properties. Up to the present time, the following properties of the near surface have been determined:

1. Wave speeds for longitudinal and transverse waves;*

2. seismic boundaries and their nature (reflecting or refracting); and

3. properties which characterize the filtering in the media, such as the response functions for transmission through and reflection from the low-speed zone and for the underlying rocks.

An example of a model for the near surface rocks is shown in Fig. 12. We will examine this model with respect to its filtering properties. The most effective filtering in the near surface rocks is done by the low-speed zone. The response function for this zone, incorporating the effects of inhomogeneities and attenuation, is a very sharp filter with a maximum at 30-40 cps.

A reflected wave arriving from below arrives at this zone complete, and the primary role in wave reflection is played by the base of the zone, with the upper surface being less important. The response function for reflection from the low-speed zone is a decaying oscillating curve.

In the medium beneath this zone, there is no significant filtering of the seismic waves, inasmuch as the rocks are nearly homogeneous and the only effect is from attenuation.

Conclusions

As a result of an examination of waves recorded at the surface and within the Earth, it appears that on transmission through the low-speed zone, the character of a wave is changed

---

*These results have been described by Gamburtsev and Koptev, presented here on p. 186.

markedly by filtering in the low-speed zone. In studying the filtering properties of the low-speed zone, it is most desirable to use direct waves recorded at the surface and at depth, or ghost events formed by reflection of waves from the low-speed zone.

For a finite number of reflections within a layer bounded by the free surface, the reflection factor depends on frequency, while the reflection factor does not depend on frequency if all possible reflections within the layer are considered. The transmission coefficient for the low-speed layer is frequency dependent.

The possibility of determining the response function for the low-speed zone from experimental data was indicated in principal. Comparison of such results with theory has shown that differences may be explained primarily by the effect of attenuation in the medium.

The response function for reflection of waves by the low-speed zone is an oscillatory curve, decaying with increasing frequency. The response function for transmission is also oscillatory; a maximum may be recognized at a frequency of 30-40 cps. The low-speed zone may be considered to be a narrow-band filter for wave transmission.

The rocks underlying the low-speed zone in the area where these studies were carried out are practically homogeneous and the only filtering is a consequence of attenuation.

We have introduced the concept of fictitious response functions. We have pointed out that these may be used in evaluating the attenuation properties of the low-speed zone and the underlying rocks.

A compilation of all these data has been used to construct a model of the upper part of the section.

LITERATURE CITED

1. E. I. Gal'perin, On the question of the effect of the surface and upper part of the section on the character and structure of a seismogram, in: Aspects of the Dynamic Theory of Propagation of Seismic Waves, No. 7, Leningrad State University, 1964.
2. A. Z. Kats, On the question of considering ground conditions in seismic microzoning, Tr. Geofiz. Inst., No. 30:157 (1955).
3. D. P. Kirnos, On the possibility of exciting resonant vibrations in an alluvial layer, Tr. Seismol. Inst. Akad. Nauk SSSR, No. 117 (1945).
4. I. S. Berzon, A. M. Epinat'eva, G. N. Pariiskaya, and S. P. Starodubrovskaya, Dynamic characteristics of seismic waves in real media, Izd. Akad. Nauk SSSR, 1962.
5. I. S. Gurvich, Seismic Exploration, Gostoptekhizdat, 1960.
6. L. L. Khudzinskii, Broadband seismic station ShPSS, Izv. Akad. Nauk SSSR, Ser. Geofiz., No. 2 (1964).
7. A. M. Epinat'eva, The study of longitudinal seismic waves propagating in some real media, Tr. Inst. Fiz. Zemli, No. 14:138 (1960).
8. I. I. Gurvich, On reflections from thin layers in seismic exploration, Prikl. Geofiz., No. 9 (1952).
9. A. A. Kharkevich, Spectra and Analysis, GITTL, 1957 [English translation: Consultants Bureau, New York, 1960].
10. L. L. Khudzinskii and A. Ya. Melamud, Equipment for frequency analysis of seismic vibrations, Izv. Akad. Nauk SSSR, Ser. Geofiz., No. 9 (1959).
11. V. V. Kuznetsov, Determining the average attenuation factors and reflection factors from the spectrums and amplitudes of direct and reflected waves, Tr. Inst. Fiz. Zemli, No. 34:201 (1964).

COMBINED STUDIES OF THE
VELOCITY CHARACTERISTICS IN THE
UPPER PART OF THE SECTION

A. G. Gamburtsev and V. I. Koptev

The question of determining the velocity characteristic of the upper part of the section is of considerable importance in seismic exploration. The upper part of the section may be studied using acoustic logging [1], well shooting, or interval velocity surveys in wells [2, 3], or by tracing of ray paths with buried geophones [4-7]. The order of preference for these methods is acoustic logging, well shooting [8-10], and then the other methods.

The present paper describes an attempt to study the velocity characteristics of the upper part of the section (over the depth range 0-80 m) using a variety of approaches. The purpose of this work was to determine the velocity profile in the near surface rocks and compare the results obtained by the various methods.

We have used data obtained by the Southern Experimental Seismic Team during 1963-1964 in the northern part of the Krasnodar Belt. The upper part of the section (from the top down) consists of the organic layer, a dry clay layer (to a depth of 6-7 m) and water saturated clays beneath.

Experimental Method and Recorded Waves

The following types of data were gathered in this study: 1) well shooting with recording using a geophone string on the surface; 2) well shooting with recording using buried geophones; 3) wave tracing between two wells; 4) vertical profiling; and 5) acoustic logging.

We will consider the nature of each of these methods.

Well Shooting with Surface Recording was done along profiles passing through the mouth of the shot hole. Vertical geophone spreads were placed 40-220 m from the shot hole, while horizontal geophone spreads were placed at 12-45 m. The spacing between geophones was greater at the greater distances, varying from 1-2 m to 10 m.

Well Shooting with Subsurface Recording (well shooting using the two-well method [10]) was done using a geophone cable hanging in drilling mud at depths of 15-80 m with a geophone spacing of 1-5 m. At shallower depths, geophones were placed in shallow drill holes. Shots were fired in a well at an offset distance of 3-5 m from the geophone hole. Vibrations were generated with blasting caps, detonated on a special shooting tool at spacings of 1 or 2 m.

Seismic Wave Tracing. The method of tracing ray paths in the medium between two wells was used in determining the wave speeds in the media beneath the base of the low-speed zone. The depth of the wells was 70 m, the spacing between them was 27 m, the geophone spacing in a well was 2.5-5 m, and the shot spacing was 2 m.

Vertical Profiling. A method devised by E. I. Gal'perin [4] was used in constructing ray paths, with several simplifying modifications; first arrivals were recorded with vertical geophones in a well. Shots were fired in wells at various distances from the recording locations (the greatest distance was 700 m). The shot depth was held constant.

The best description of the method may be found in the paper by Kuznetsov and Gamburtsev presented on p. 163.

Used were seismic recorders PMZ-1, ShPSS-24, and ShPSS-48 [11], with SPM-16 geophones.

Acoustic Logging. A description of the acoustic logging equipment may be found in reference [1].

Brief Description of the Results Obtained

The following waves were recognized in the early parts of the records: direct longitudinal and transverse waves P_1 and S_1, a ghost of the direct wave, P_1^g, formed by reflection of the direct wave in the low-speed zone, and what seem to be head waves to the boundary between the low-speed zone and the underlying rocks.*

The characteristics of waves recorded from shots below and within the low-speed zone are as follows.

With the shot point beneath the low-speed zone at the depth of the water table, the first arrival that can be recognized is the direct wave. The apparent frequencies range from 80 cps (for observations at the surface) to 200 cps (for observations in wells). With increasing depth of burial for the geophones, the frequency increased. As a result, the records obtained with the deeper geophones have sharper first arrivals than those obtained at shallow depths or on the surface.

The vertical geophones located in wells detected the direct transverse waves well. The direct transverse wave can be traced on the surface horizontal geophones practically from the mouth of the shooting hole (x = 1-1.5 m).

Examples of records obtained for shots beneath the low-speed zone are shown in Fig. 1. On the left are waves recorded on vertical geophones located on the surface and at various depths in wells; to the right are waves recorded with horizontal geophones located on the surface (for a shot of one detonator at 25-m depth); the spacing between geophones is 2.5 m; P_1 is the direct longitudinal wave, P_1^g is the ghost of P_1, and S_1 is the direct transverse wave.

With movement of the shot point through the bottom of the low-speed zone, the picture changes markedly. Head waves related to the rocks beneath the low-speed zone are clearly apparent in the vicinity of the first arrivals on records from the geophone spreads (Fig. 2). These waves are relatively weak and quickly die out with greater distance from the source. It is possible to separate the direct longitudinal wave from the head waves in the later part of the records.

In Fig. 2 the spacing between geophones is 2 m. P_{pr} is a refracted wave from rocks beneath the low-speed zone; a is a shot point beneath the low-speed zone, h = 20 m; b is a shot point in the low-speed layer, h = 18 m.

---

*It is possible that these are not head waves, but refracted waves from the underlying rocks.

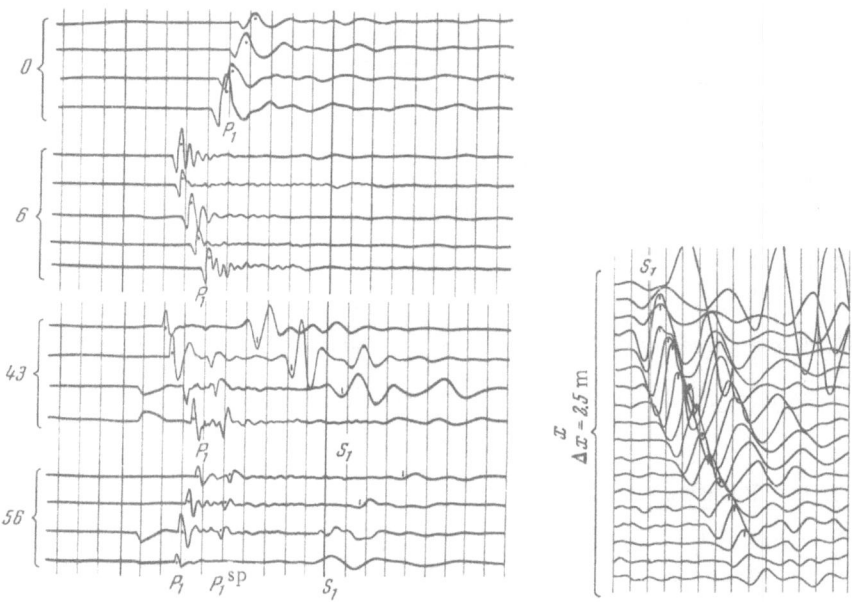

Fig. 1. Examples of records obtained in well shooting.

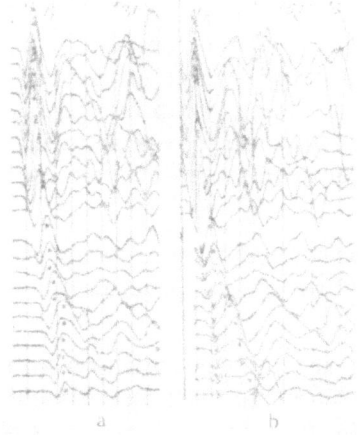

Fig. 2. Records from geo-
phones located along a surface profile, with the shot
point located in or below the
low-speed zone.

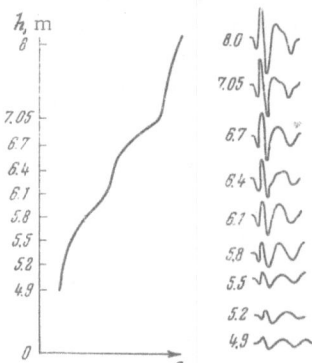

Fig. 3. Seismic traces and
amplitude curve showing the
change in amplitude of direct
waves as the shot point is
moved across the base of the
low-speed zone.

Geophones located beneath the bottom of the low-speed zone continued to record a direct
longitudinal wave, which decreased quickly in amplitude as the shot point was moved across the
base of the low-speed zone. A group of first arrivals is shown in Fig. 3 for a geophone located
at a depth of 20 m, with the shot point being moved at intervals of 30 cm with the same gain on
all traces. It is apparent from this illustration that the amplitude of the direct wave changes
only slightly, by a factor of 1.15, over the depth interval from 8-7.05 m, but it changes by a

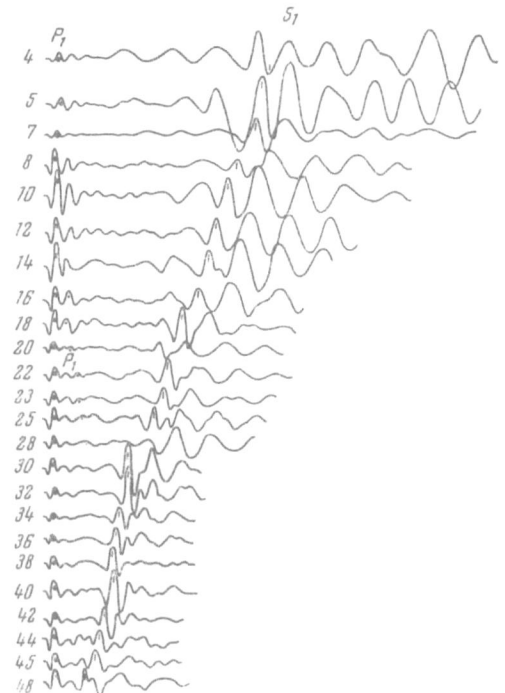

Fig. 4. Seismic traces showing the character of the direct longitudinal and transverse waves for shots at various depths.

factor of 2 over the range 7.05-6.1 m, and by 2.3 over the range 6.1-5.2 m. The base of the low-speed zone is at 6-7 m depth. Thus, despite the strong gradual decrease in amplitude, it is possible to relate the amplitude of the wave to the properties of the medium, and to determine where the base of the low-speed layer is from the change in slope. Moreover, it is apparent that the base of the low-speed zone cannot be considered to be a sharp boundary. It is a gradual transition from dry rocks to water-saturated rocks. Physically, this apparently corresponds to the zone of capillary saturation seen in wells (see, for example, reference [12]). It is possible that there is also a discontinuous change in the properties of the medium at the base of this region.

A group of records obtained over a large range of shot depths (with a geophone located at a depth of 66 m) is shown in Fig. 4. There is no significant change in the form of the direct longitudinal wave over a range of shot depths from 48 to 16-20 m. When the shot points are close to the low-speed zone (with $h_{shot} < 20$ m), the direct wave begins to overlay its own ghost P_1^g, formed at the base of the zone. At the same time, the direct transverse wave, which is less stable in form, does not change significantly at the base of the low-speed zone (its amplitude increases rapidly for shots at depths less than 5 m, but this is a consequence of the behavior of the amplifier).

It follows from this description of the character of the direct waves that the medium beneath the low-speed zone is comparatively homogeneous, inasmuch as the form of the longitudinal direct wave does not change with shot depth. It is possible that this part of the medium is not so uniform with respect to transmission of shear waves. The low-speed zone appears to have a lower boundary which provides sharp reflection and reflection of longitudinal waves, but not of shear waves. The depth to the base of the low-speed zone may be determined easily from the appearance of head waves on the geophone spreads and from the strong change in amplitude of the direct waves at geophones in drill holes.

In the acoustic logging, a good correlation of the records of the longitudinal waves with an apparent frequency of about 50 kcps was obtained. Good results could be obtained only below the low-speed zone; that is, in rocks located below the water table.

Data on Wave Speeds in the Near-Surface Rocks

Acoustic logging data, well shooting data, wave tracing data, and vertical profiling data were all used in constructing a velocity profile. The combined use of all of these data results in a very complete representation of the velocity profile and of the seismic boundaries in a region under study.

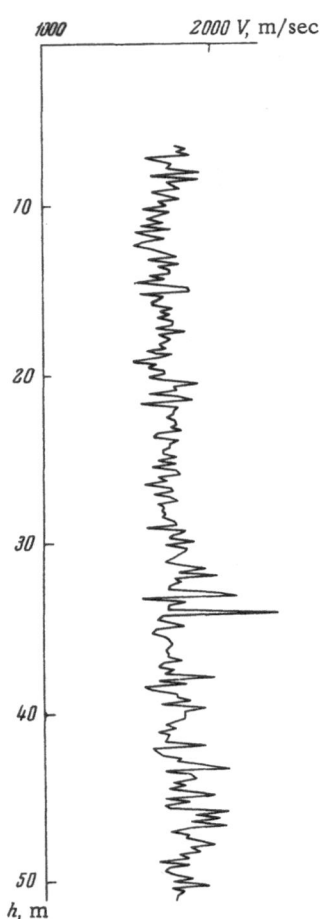

Fig. 5. Velocity profile constructed from acoustic logging data.

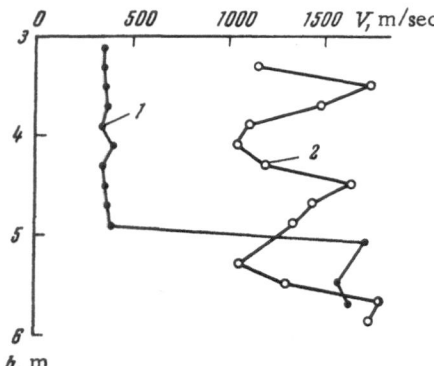

Fig. 6. Acoustic log obtained within the low-speed zone in a dry hole (1) and in a hole filled with drilling mud (2).

Velocity Profile Obtained from Logging Data and Well Shooting (Normal Incidence)

Logging Data. An acoustic wave speed log from a well in the region under consideration is shown in Fig. 5. It is seen that the rocks beneath the low-speed zone are thin bedded with minor changes in wave speed.* The average gradient of wave speed with depth over the whole interval studied is about 1 m/sec/m.

Data from repeat surveys indicate that the standard deviation for wave-speed measurements made over a span of 20 cm is 90 m/sec, or 5%.

The data show that no layers thicker than 1-2 m can be recognized in the section.

Acoustic logs were run within the low-speed zone using a dry hole and a hole filled with drilling mud. The velocity profiles which were obtained are shown in Fig. 6. Obviously, one of these is not correct. The wave speeds determined in the drill hole filled with mud are apparently too high, and the points show considerable scatter because the waves are propagating in rocks with variable water content. The wave speeds obtained in the dry hole are too low in comparison with the more reliable wave speeds found from well shooting. It is possible that the broken zone around the drill hole is detected with the acoustic log in this case.

Thus, a method for acoustic logging in dry rocks with low wave speeds needs to be developed.

Determination of a Velocity Profile from Well Shooting. A method for recording and interpreting well shooting data from a well (or wells) offset from a shot hole by a very small distance (3-5 m) has been described in detail in reference [10]. Direct waves are recorded with geophones located above and below the shot points. We have called this the "two-well well shooting method."

The main points of the method of interpretation are as follows. For each interval Δh between shot depths, the value for the interval time τ, which is the transit time for waves between shot points,

_____

* The sonic logs from neighboring wells were generally the same.

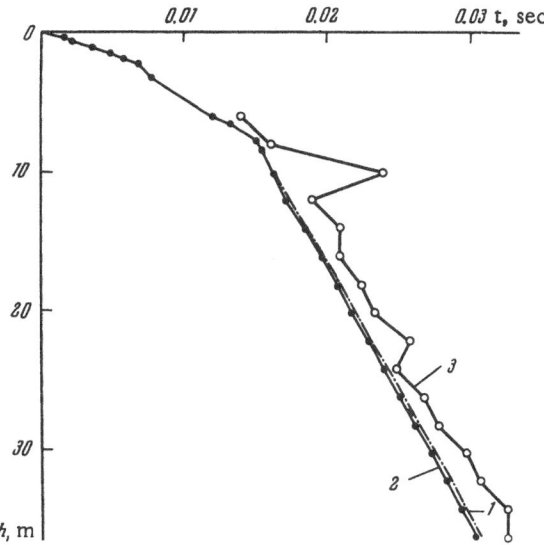

Fig. 7. Vertical travel-time curves constructed from acoustic logging data (1), the two-well well-shooting method (2), and from data from a single geophone located at the mouth of the shot hole (3).

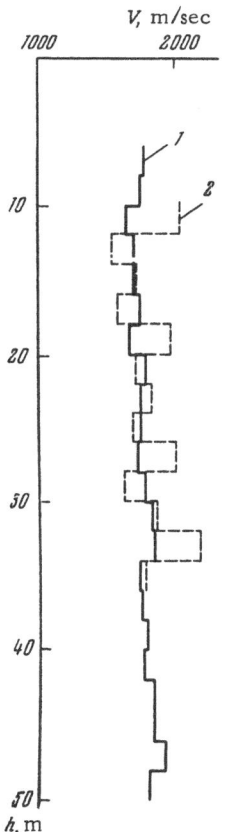

Fig. 8. Interval velocity curves obtained from an acoustic log and a well shooting survey. 1) Logging data; 2) shooting data.

is determined. The value for τ is determined from the difference in travel time for waves from two adjacent shot points, using a fixed geophone. In each case, the value for τ is determined with some total error which includes various random errors related to the accuracy of time measurements on the seismograms, the effect of minor inhomogeneities on propagation times, and so on, as well as systematic errors caused by inaccurate determination of the shot instant on the records. With the determination of time differences from geophones located above and below the shot point and an arithmetic average of $\bar{\tau}\!\uparrow$ and $\bar{\tau}\!\downarrow$ (where $\bar{\tau}\!\uparrow$ and $\bar{\tau}\!\downarrow$ indicate time differences obtained with only a geophone above the shot points and only a geophone below the shot points, respectively, for a given pair of shot points), the random errors are reduced by a factor $\sqrt{n}$, where n is the number of geophones located above or below the shot points. The systematic errors caused by the incorrect shot instant remain, in these averages. The error enters $\bar{\tau}\!\uparrow$ and $\bar{\tau}\!\downarrow$ with opposite signs, therefore, the arithmetic average, $\bar{\tau} = (\bar{\tau}\!\uparrow + \bar{\tau}\!\downarrow)/2$ does not contain this systematic error.

Further analysis of the interval times may be done by various means (for example, see reference [9]). As will be shown later, the use of this approach to interpreting data will increase the precision with which the velocity profile may be constructed significantly in comparison with the usual method of well shooting, where only records made at the surface are used.

Well Shooting Data and Their Comparison with Well Logging Data. The vertical travel–time curve shown in Fig. 7 was constructed from the values for interval times obtained from shooting below the base of the low-speed zone. As is apparent from this illustration, the vertical travel–time curve is nearly a straight line, and the rocks beneath the low-speed zone may be considered to be practically homogeneous. The vertical travel–time curve constructed from data from a single geophone is broken, and does not give a correct impression of the velocity profile. The vertical travel–time curve obtained from acoustic logging data is also shown on this illustration for comparison. It is very close to

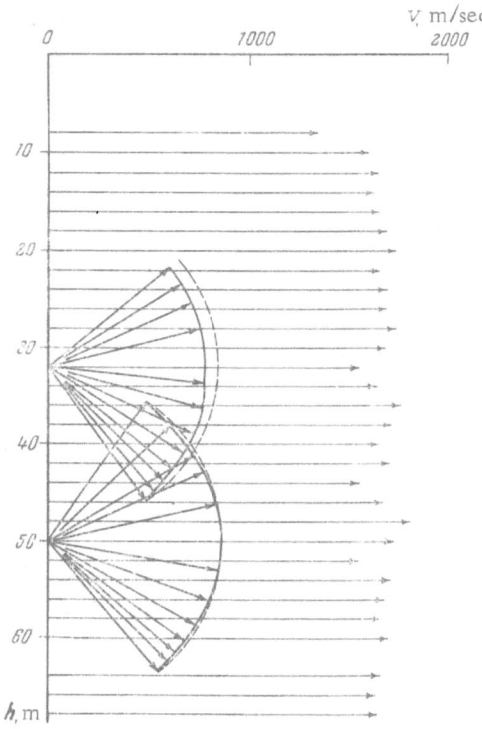

Fig. 9. Graphs illustrating the character of the velocity V_P in the horizontal direction and the velocity indicatrix (the latter data are on a reduced scale).

the travel−time curve obtained from well shooting data in form and slope.

Curves for interval velocities determined from acoustic logging data and from well shooting are shown in Fig. 8. The velocity over 2 m intervals was determined in both cases. It is apparent that the interval velocities determined from logging data show less scatter than the values determined from well shooting. The disagreement between interval velocities reflects primarily the difference in precision characterizing the logging and well shooting.* A statistical analysis of the data does not permit the subdivision of the section beneath the low-speed zone into recognizable zones.

Reduction of vertical travel−time curves constructed from data observed in the low-speed zone (a group of travel−time curves, one of which was shown in Fig. 7, was used) indicates that the low-speed zone has a gradient in wave speed. The wave speed varies almost linearly with depth, at a rate of about 30 m/sec/m. Near the base of the low-speed zone, the gradient increases to 100-300 m/sec/m, but the transitional region is not sufficiently well known, and a definite description of the boundary cannot be given.

Thus, the best method for studying wave speeds beneath the low-speed zone is the acoustic logging method. It provides the most accurate and detailed information about the velocity profile. Comparison of acoustic logging data with well shooting data indicates that results obtained at widely different frequencies provides the same velocity profile, though with different precisions and degrees of detail.

No good method is presently available for determining wave speeds in the low-speed zone. The well shooting method does not provide adequate precision and detail for subdividing the zone.

Determination of the Velocity Profile from Shear Wave Data in the Near-Surface Rocks

It was shown earlier (Fig. 1) that good direct shear waves were recorded on vertical geophones in wells with firing of a detonator. Using a number of geophones which are reasonably far away vertically from the shot level, it is possible to determine interval velocities and to construct a vertical travel−time curve accurately from time differences for travel times of shear waves between adjacent geophones.

The vertical travel−time curve constructed for the S_1 wave is smooth, becoming steeper with increasing depth. The gradient is quite apparent, and the gradient decreases quickly with depth. Discontinuities in wave speed, if present, are not large.

In constructing a vertical travel−time curve from data recorded at the surface, with shallow shot depths, the method of squared coordinates [13] applicable to direct waves may be

---

*Estimates for specific materials, from all types of observations, are given in [10], and by Kondrat'ev and Saks in this book.

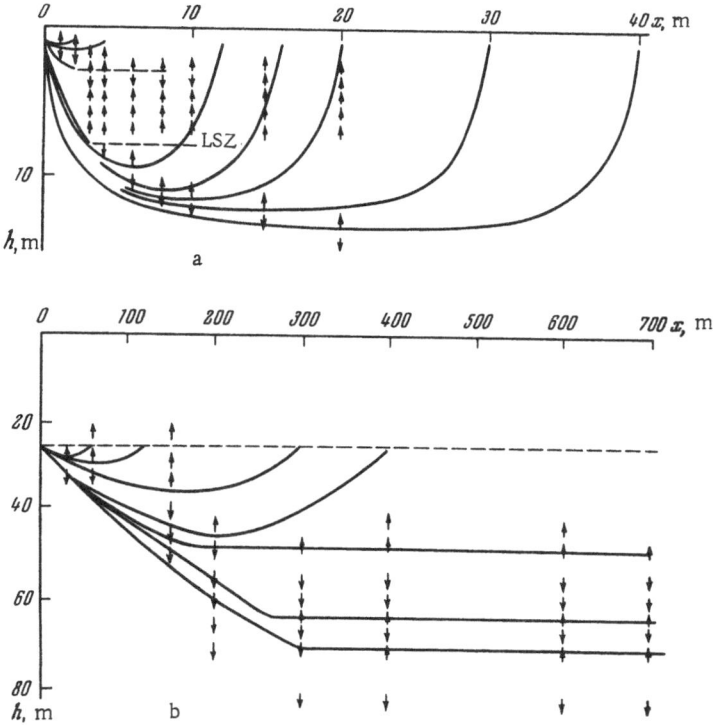

Fig. 10. Ray paths for first arrivals in depth intervals.
a) 1-5 m; b) 25-80 m; LSZ — low-speed zone.

The maximum depth of penetration for a refracted-ray path is found on the vertical profiles between geophones having opposite directions for first arrivals (the upper geophone indicates an upward vector component, while a lower geophone indicates a downward vector component). If, beginning at some distance from the "source" the change in direction is observed at the same depth in all wells, the presence of a refracting boundary or thin refracting layer may be assumed.

A ray path system obtained in this way is shown in Fig. 10. This illustration indicates that: 1) there is a refracting boundary at a depth of 2-3 m in the low-speed zone; 2) the base of the low-speed zone is a very sharp refracting boundary; 3) there are weak refracting boundaries at depths of 48, 64, and 70 m; and 4) the medium beneath the low-speed zone shows only minor changes in wave speed and is characterized by a slight gradient in wave speed; a gradient of about 1 m/sec/m was calculated using the expression given in [13].

The locations of isochrons constructed from surface observations and well shooting (see, for example, [3]) confirm the first two conclusions.

Representation of the Velocity Profile

A combination of the various data obtained in the regional work leads to a representation for the velocity profile in the near-surface rocks.

In the region under consideration, the near-surface rocks include a low-speed zone 6-7 m thick and the underlying rocks (Fig. 11).

The low-speed zone is an inhomogeneous medium with a single refracting boundary at 2-3 m depth. Wave speed for longitudinal waves increases with depth from 270 to 1100 m/sec. The wave speed increases abruptly across the base of the low-speed zone. The base of the zone is a strong reflecting and refracting boundary for longitudinal waves. The rocks beneath the low-speed zone show a weak gradient; they are nearly homogeneous in structure, but have refracting boundaries. The model for shear waves is quite different. The low-speed zone has no effect on their wave speeds. The velocity profile for shear waves may be represented in terms of gradients. The gradients of shear wave speeds decrease with depth. Similar results have been obtained and described in "The Construction of a Physical Model of Actual Media and the Study of the Corresponding Wave-Field Effects," published by Fondi Inst. Fizika Zemli, Akad. Nauk SSSR, 1964.

The value for $\bar{1}/\gamma = V_P/V_S$ varies from 1.7-2 (in the low-speed zone) to 4-6 (in the underlying rocks).

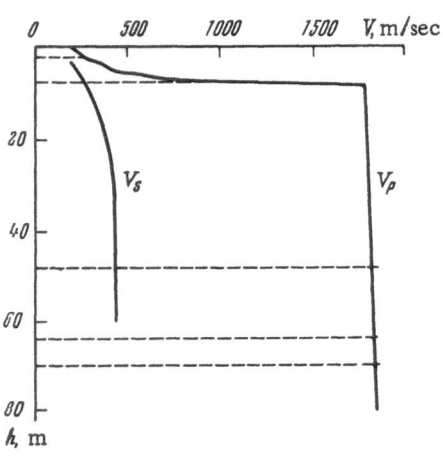

Fig. 11. Velocity profiles.

Conclusions

The advantages of using a variety of data for studying the velocity profile in the upper part of the section have been demonstrated in this study. The most accurate information on longitudinal-wave speeds in the rocks beneath the low-speed zone is provided by acoustic logging. Data obtained at ultrasonic frequencies generally agree quite well with measurements made at seismic frequencies (well shooting).

Information on longitudinal wave speeds in the low-speed zone, and on shear wave speeds for the whole section was obtained from well shooting. However, well shooting cannot be accepted as the best method for determining these wave speeds, and an attempt should be made to improve the precision with which they are determined.

In observing emergence angles for direct waves, it was possible to recognize several refracting boundaries and to determine longitudinal wave speeds as a function of direction.

This use of combined data provides a highly satisfactory description of the velocity profile.

LITERATURE CITED

1. V. I. Koptev, Study of elastic properties, mechanisms for ground compaction, and the degree of competence of fractured rocks by ultrasonic methods, Tr. Gidroproekta, No. 9 (1963).
2. N. N. Puzirev, Measurement of seismic velocities in wells, Tr. Vses. Nauchn.-Issled. Inst. Geofiz., Issue III (1957).
3. I. S. Berzon, Some results of the study of seismic waves with well shooting, Izv. Akad. Nauk SSSR, Ser. Geofiz., No. 9 (1964).
4. E. I. Gal'perin, Description of a detailed study of velocity models of the upper part of the section under conditions of weak velocity differentiation, Izv. Akad. Nauk SSSR, Ser. Geofiz., No. 4 (1964).
5. V. V. Zhadin, Investigation of the decay and dispersion of seismic waves by well surveys, in: Aspects of the Dynamic Theory of Seismic Wave Propagation, No. 2, Leningrad State University, 1959.
6. N. I. Berdennikova, V. V. Zhadin, and A. G. Rudakov, On the question of methods for seismic logging, in: Aspects of the Dynamic Theory of Seismic Wave Propagation, No. 2, Leningrad State University, 1959.
7. N. I. Berdennikova, On some effects of anisotropy in layered media for work with shear waves, in: Aspects of the Dynamic Theory of Seismic Wave Propagation, No. 2, Leningrad State University, 1959.
8. Yu. B. Demidenko, Seismic logging of shot holes. Razved. i Promis. Geofiz., No. 30 (1959).
9. V. M. Glogovskii, On statistical methods for solving some problems in the reduction of seismic travel−time curves, Geofiz. Razvedka, No. 7 (1962).
10. A. G. Gamburtsev, Method of determining the velocity characteristic of a section with the aid of well shooting with observations at interior points of the medium, Prikl. Geofiz., No. 49 (1967).
11. L. L. Khudzinskii, Broadband seismic station ShPSS, Izv. Akad. Nauk SSSR, Ser. Geofiz., No. 4 (1963).
12. N. A. Tsitovich, Soil Mechanics, Stroiizdat, 1940.
13. G. A. Gamburtsev, Fundamentals of seismic prospecting, Gostoptekhizdat, 1959.

SECONDARY PULSES ON RECORDS OF
DIRECT WAVES FROM SHOTS IN WELLS

V. V. Kuznetsov and Yu. A. Ostrovskii

In recent years, much attention has been paid to the study of the nature and characteristics of the propagation of direct waves in real media [1-4]. This is because the form of the seismic pulse is required in the solution of a number of practical problems. Among these problems are those of controlling the shooting conditions, of determining the properties of the medium (the frequency dependence of the attenuation and reflection factors, and so on), of obtaining source data for constructing synthetic seismograms, and so on.

Direct observation of the incident pulse at interior points of the medium is subject to a number of difficulties. Therefore, it is preferable to record the direct waves at the surface and to use the recorded pulse as the incident pulse. The basis of this approach has been described in the literature (see, for example, references [3-4]). However, at times not one but several pulses are recorded as the direct wave at the surface. These were termed secondary emissions in reference [5]. The nature of these emissions was not explained. It has been suggested that the secondary emissions are caused by a gaseous-bubble pulse in the shot hole [6, 7]. This indicates the need for a study of the probable causes of the secondary pulses seen on records of the direct wave.

In setting up the experimental and theoretical studies, the initial hypotheses about the origin of the secondary pulses were:

1. The formation of a secondary pulse on a record of the direct wave is the result of a pulsation in gas density formed in shooting in a well filled with water;

2. the secondary pulse is formed as a result of distortion of the direct wave by the recording equipment for large displacements of the free surface; or

3. the secondary pulse is formed by conversion of a parabolic displacement of the free surface by the seismic transducer.

In this paper, we examine the possibility of explaining the secondary pulses on records of direct waves by each of these hypotheses, and draw conclusions concerning the origin of these pulses.

Experimental Conditions

Equipment. In the study of the origin of secondary pulses, we used PMZ and ShPSS [8] equipment. We used electrodynamic geophones of type SPM-16, with a frequency of 32 cps, and galvanometers with a frequency of 130 cps.

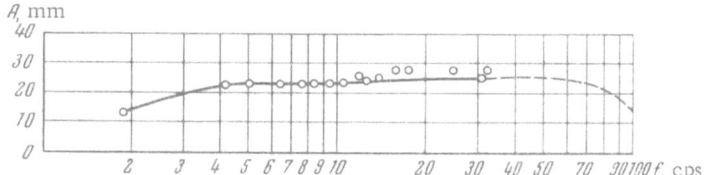

Fig. 1. Frequency response for the VEGIK seismograph and GB-IV galvanometer. Solid line — observed response; dashed line — calculated response.

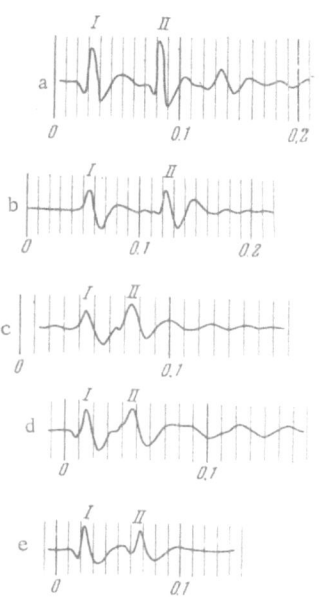

Fig. 2. Group of records of the primary (I) and the secondary (II) pulses made at the surface. a) $h_{shot} = 30$ m; $Q = 20$ kg; b) $h_{shot} = 26$ m; $Q = 10$ kg; c) $h_{shot} = 15$ m; $Q = 2.5$ kg; d) $h_{shot} = 10$ m; $Q = 0.3$ kg; e) $h_{shot} = 5$ m; $Q = 0.3$ kg.

We used the following types of geophones in evaluating the possibility of explaining the secondary pulses as the distorting effect of the recording equipment: SP-15 ($f = 15$ cps), ASED ($f = 12$ cps), and SEDS ($f = 10 \sim 12$ cps), as well as piezoelectric transducers of type TsTS-19.

With the PMZ equipment, in recording direct and reflected waves simultaneously, we used broadband recording (passband from 60 to 400 cps), a high-frequency setting ($f = 135$ cps) and a mid-frequency setting ($f = 45$ cps). Direct recording was used with the PMZ equipment.

With the ShPSS equipment, in recording the direct and reflected waves, broadband response was used, with a much wider passband than in the PMZ equipment. With the use of a electrodynamic geophones, the recorded response is the amplitude of geophone case velocity over a frequency range from just above the geophone frequency to just below the galvanometer frequency.

The output of the seismic amplifiers was divided down with a resistive network to provide readable records of the direct waves.

A type VEGIK [9] seismograph connected to a type GB-IV galvanometer (the seismograph period was 1 sec and the galvanometer frequency was 15 cps) was used to record the displacement of the free surface. The frequency response of the seismograph—galvanometer system is shown on Fig. 1. Records were made on an H-700 oscillograph [10].

Well Section. The structure of the upper part of the section in which the shot holes and geophone holes were located differed at different survey sites.

Studies were carried out in the Genichesk area of the Kherson region (location I), in the Dzhankoi area of the Simferopol region (location II), and in the Leningrad area of the Krasnodar Belt (location III). The characteristics of these locations were as follows: a uniform shale with a longitudinal wave speed of 1700-1800 m/sec lies beneath the low-speed zone, the rocks are characterized by low attenuation, and the base of the low-speed zone is a strong reflecting and refracting boundary.

The structure of the low-speed zone varies between the different locations. At location I, the zone consists of two layers with a thickness of about 15 m. The wave speed in the upper layer is 300-350 m/sec, and in the lower layer, 500-600 m/sec. At location II, the low-speed

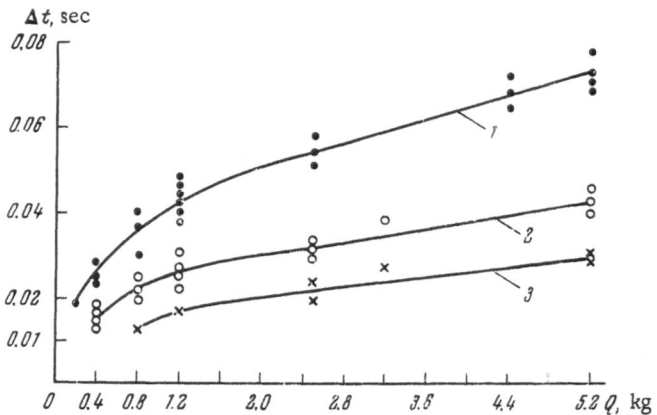

Fig. 3. Curves for the relationship of time difference Δt between primary and secondary pulses to charge size. 1) $h_{shot} = 10$ m; 2) $h_{shot} = 20$ m; 3) $h_{shot} = 30$ m.

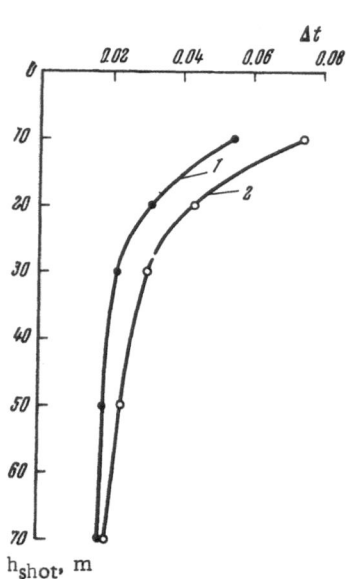

Fig. 4. Curves for the relationship of time difference Δt between primary and secondary pulses to shot depth. 1) $Q = 2.5$ kg; 2) $Q = 5.2$ kg.

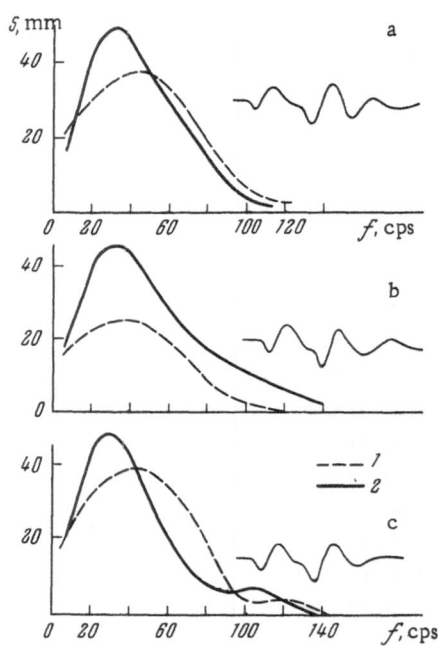

Fig. 5. Spectrums for the primary (1) and secondary (2) pulses at different recording points. a) $h_{shot} = 30$ m, $Q = 2.5$ kg; b) $h_{shot} = 30$ m, $Q = 1.2$ kg; c) $h_{shot} = 25$ m, $Q = 1.2$ kg.

zone is a single layer with a thickness of less than 5 m, with a longitudinal wave speed of 500-550 m/sec. At location III, the low-speed zone is inhomogeneous with a thickness of about 6 m, with one refracting boundary at a depth of 2-3 m. The longitudinal wave speed increases with depth from 270 to 700 m/sec.

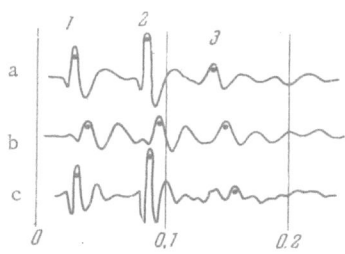

Fig. 6. Group of records of the primary (1), secondary (2), and tertiary (3) pulses recorded with PMZ equipment. a) Broadband filtering; b) mid-frequency passband; c) high-frequency passband; h_{shot} = 30 m, Q = 20 kg.

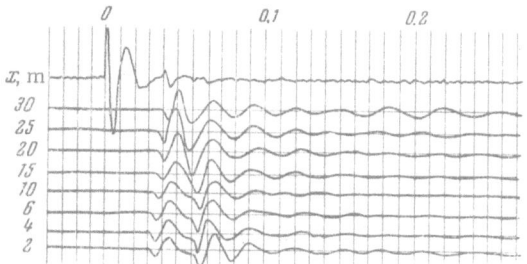

Fig. 7. Example of traces with direct waves recorded along a profile with vertical geophones. h_{shot} = 30 m; Q = 3.1 kg.

Shots at all three locations were fired in the uniform shale beneath the low-speed zone.

General Relationships. The form of the pulse generated by a shot in a well is most clearly apparent on the record of direct waves made from the geophone located at the mouth of the shot hole. When secondary pulses are absent, the record of the direct wave is short and simple in form [3, 5]. With increasing charge size or decreasing shot depth, the direct wave pulse becomes double (Fig. 2). The time difference Δt between arrival of the primary and secondary pulses increases with increasing charge size and decreasing shot depth. Curves giving the relationship between Δt and charge size for fixed shot depths are given in Fig. 3. The empirical relationship $\Delta t = f(Q)$ may be expressed approximately with the formula $\Delta t = k\sqrt[3]{Q}$. Curves for the relationship of Δt and shot depth for fixed charge sizes are shown in Fig. 4. A smooth decrease in Δt with increasing depth of the shot is observed.

The forms of the primary and secondary pulses are similar. The ratio of amplitudes is close to unity. In some cases, the amplitude of the secondary pulse is somewhat larger than that of the primary. This relationship is most clearly apparent in cases in which the two pulses are not completely separated in time.

The spectrum of the secondary pulse is shifted towards lower frequencies and is narrower than the spectrum of the primary pulse (Fig. 5). The shift in the maximum is not large, amounting to 5-10 cps.

Three pulses were recorded in some of the experiments (Fig. 6). The amplitude of the third pulse is much smaller than that of the first two. The ratio of the amplitude of the third pulse to the amplitude of the first two is less than unity and changes with filter settings.

The time difference between the arrivals of the second and third pulses is less than the time difference between the first two.

The strongest third pulses were recorded with mid-frequency passbands and the weakest with high-frequency passbands. Thus, the third pulse has a lower frequency spectrum than the first two.

We see that some of the characteristics of the secondary pulses on records of direct waves are the same for shots in wells as for shots in water (the increase in Δt with increasing charge weight, the decrease in amplitude and time difference with higher order of pulse). However, the absolute sizes for time differences between the primary and secondary pulses for shots in wells are much less than for shots in water. They differ by a factor of 5 to 8 for a given shot depth and charge size [6, 7].

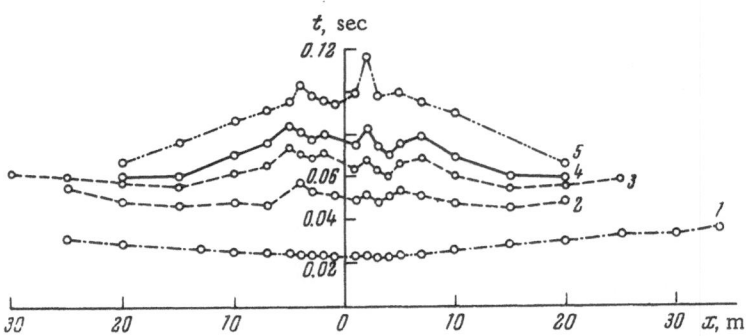

Fig. 8. Travel—time curves for the first arrival (1) and later phases (2-5) of the direct wave for various size shots at a fixed depth of 10 m. 2) $Q = 0.8$ kg; 3) $Q = 1.2$ kg; 4) $Q = 2.5$ kg; 5) $Q = 10$ kg.

Pulsation of Gas Density in the Shot Hole as an Explanation of the Secondary Pulse

We measured the following factors to test the hypothesis about the generation of the secondary pulse as a consequence of the pulsation of a gas bubble:

1. The region of propagation was determined for the secondary pulses with surface observations, as well as at interior points of the medium;

2. the effect of the secondary pulses on the reflected wave forms was determined; and

3. the shooting conditions were changed markedly (shots were fired in dry holes).

Range of Propagation. If it is assumed that the secondary pulses on records of direct waves are caused by a pulsating gas bubble, the time difference between the primary and secondary pulses should be the same at all distances, whether records are made at the surface or at interior points of the medium.

For this reason, direct waves were recorded at surface points along profiles passing through the mouth of the shot hole and at interior points of the medium at a variety of distances.

An example of a record of direct waves recorded along a horizontal profile is shown in Fig. 7. The profile length is 30 m. Considering this record, it is apparent that Δt decreases with increasing distance from the mouth of the shot hole, and at the greater distances from the shot point, with $l = 43$ m ($x = 30$ m), a single pulse is recorded (Δt is zero in this case).

Travel—time curves for the first arrival and the later phases of the direct waves recorded with various charge sizes at one shot depth are shown in Fig. 8. With increasing charge size, the travel—time curves for the later phases are found not to be parallel to the curve for the first arrival. The amount of nonparallelism decreases with increasing distance from the mouth of the shot hole, and at some distance, the travel—time curves become parallel. The behavior of the travel—time curves for the later phases is caused by the fact that Δt decreases with distance from the source.

Thus, it is seen that there is a limited range for observation of the secondary pulses on records of the direct waves made at the surface.

Measurements were made within the medium to improve the reliability of the results. Geophones to detect the direct waves were placed in a ring of drill holes with a radius of 10 m. Shots were fired with various charge sizes at four fixed depths (10, 20, 30, and 50 m) in a well

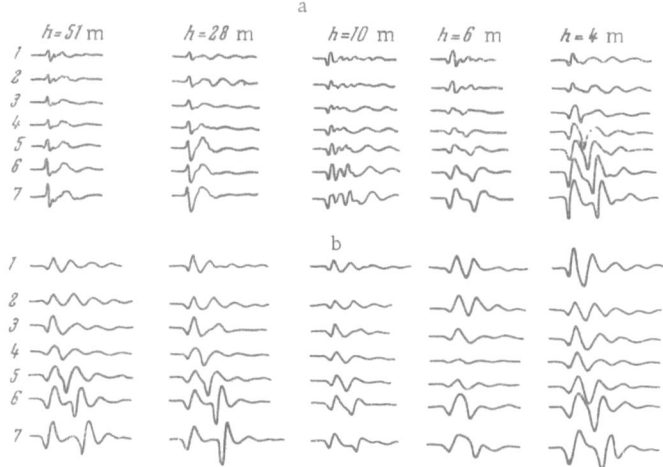

Fig. 9. Groups of records of direct waves recorded at interior points of the medium with various geophone depths (a) and at the surface at the mouth of the shot hole (b) with a shot depth of 20 m but various charge sizes. 1) 1) $Q = 10$ detonators; 2) $Q = 20$ detonators; 3) $Q = 0.4$ kg; 4) $Q = 0.8$ kg; 5) $Q = 1.2$ kg; 6) $Q = 2.5$ kg; 7) $Q = 5$ kg.

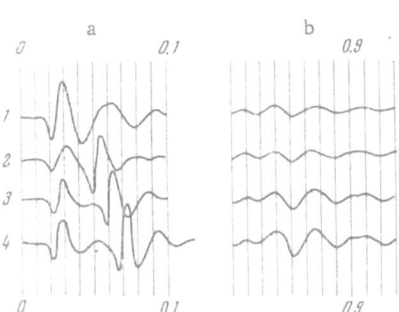

Fig. 10. Group of records of direct (a) and reflected (b) waves recorded with ShPSS equipment, for various charge sizes at a depth of 20 m. 1) $Q = 0.4$ kg; 2) $Q = 1.2$ kg; 3) $Q = 2.5$ kg; 4) $Q = 5$ kg.

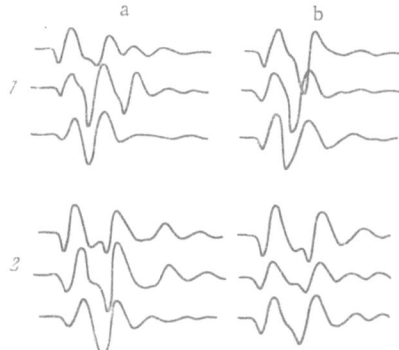

Fig. 11. A group of records of direct waves recorded with ShPSS equipment for shots in dry holes (a) with a shot depth of 6 m, and in water-filled holes (b) with a shot depth of 8 m. 1) $Q = 0.4$ kg; 2) $Q = 1.2$ kg; geophone spacing, 5 m.

drilled at the center of the ring. A single geophone was placed at the mouth of the shot hole to record the direct wave.

A group of records of the direct waves recorded in the medium and at the surface is shown in Fig. 9. The secondary pulse is absent from records of geophones buried deeper than 6 m. With small geophone depths ($h = 6$ and 4 m), the secondary pulses are clearly apparent. The surface geophone located at the mouth of the shot hole also shows the secondary pulse. The time difference between the arrivals of the primary and secondary pulses at an interior point is less than that at the surface.

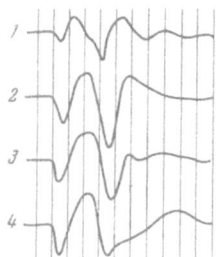

Fig. 12. Group of records of the direct wave obtained for a charge size of 2.8 kg at a depth of 22 m with different types of geophones. 1) SPM-16; 2) SP-15; 3) ASED; 4) SEDS.

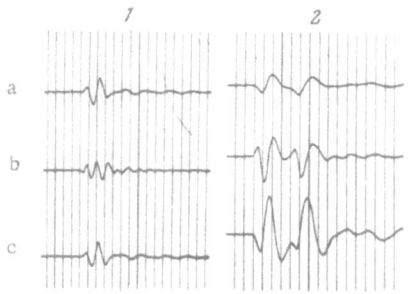

Fig. 13. Traces of direct wave recorded with the system. SPM-16 geophone-galvanometer (a), piezo-electric transducer/PMZ recorder (b), and SPM-16 geophone/PMZ recorder (c). 1) h_{shot} = 30 m, Q = 0.4 kg; 2) h_{shot} = 25 m, Q = 5.0 kg.

At the locations where these measurements were made, the thickness of the low-speed zone is no more than 6 m. Thus, clear secondary pulses are recorded within the low-speed zone and at the surface.

The limited range for the existence of the secondary pulses at interior points of the medium is confirmed by a comparison of reflected waves recorded with the presence and absence of the secondary pulse with the direct wave (Fig. 10). With the appearance of the secondary pulse the form of the reflected wave remains the same with increasing charge size, while the direct wave changes markedly. Thus, the secondary pulse has no effect on the record of a reflected wave. More detailed studies have also confirmed this result.*

The studies which have been made indicate that the secondary pulse on records of the direct wave have a limited range of propagation both along the surface and to points within the medium. These results show that the secondary pulse is not a bubble pulse from shooting in a well.

Secondary Pulses at the Surface with Shooting in Dry Holes. Direct waves were recorded from shots fired in dry holes, with the purpose of studying the characteristics of the secondary pulse for a variety of shooting conditions (Fig. 11). The records which were obtained were compared with those of the direct waves recorded at the surface from shots fired in water-filled holes. The records are similar. A secondary pulse is clearly apparent with the direct waves recorded from the shots in the dry holes. An increase in the Δt betweeen the primary and secondary pulses is seen with increasing charge weight. The values found for Δt are close to those recorded with shots in water-filled holes. The values for Δt decreases to zero with increasing distance from the mouth of the shot hole.

Thus, the secondary pulses are not caused by a bubble pulse.

Distortion in the Recording Equipment as an Explanation

of the Secondary Pulse

With measurements close to the shot point, where displacement is significant, it might be assumed that the secondary pulse is the result of distortion in the geophone-amplifier galvanometer circuit.

The greatest distortion takes place in the geophone, inasmuch as it has limited coil movement under large displacements.

---

*The question of the effect of the secondary pulse on reflected waves was considered in a paper by T. N. Ershova, V. V. Kuznetsov, and Yu. A. Ostrovskii, "Secondary emissions from shots in wells," Fondi Inst. Zemli, Akad. Nauk SSSR, 1964.

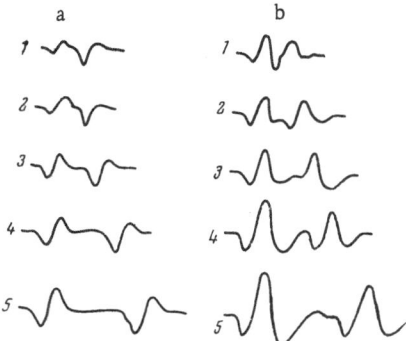

Fig. 14. Group of traces of secondary pulses recorded with an
SPM-16 geophone connected directly to a galvanometer (a),
and with PMZ broadband filters
included in the circuit (b),
with a shot depth of 24 m. 1)
Q = 0.6 kg; 2) Q = 1.2 kg; 3) Q = 5.6
kg; 4) Q = 10.0 kg; 5) Q = 12.0 kg.

In order to test this hypothesis, we recorded secondary pulses with various types of geophones. A group
of traces recorded with type SPM 16, ASED, SP 15, and
SEDS geophones is shown in Fig. 12. The forms of the
records differ markedly, but all show the secondary
pulse. The time difference between the arrivals of the
primary and secondary pulses is practically the same
for all records ($\Delta t = 27$ or 30 msec), though the permissible coil movement is quite different in the various
geophones (from 1.8 mm in the SPM-16 to 8 mm in the
SEDS). However, the measurements which were made
did not positively establish that these geophones did not
distort the records for the layer displacements, inasmuch as the amplitudes of surface motion may have ex-
ceeded the ranges for the geophones. So, the direct
waves were recorded with piezoelectric transducers,
and these records are shown in comparison with those
from an SPM-16 geophone in Fig. 13. It may be seen
that the record obtained with the piezoelectric transducer shows the same secondary pulses as the record
obtained with the SPM-16 geophone. Consequently, the secondary pulse is not caused by distortion in the geophone.

The amplifier in the seismic recorder may be another source of distortion in the rec-
cording system. The direct wave was recorded with a geophone connected directly to a galvanometer to avoid this possibility. A group of records of secondary pulses recorded in this
way is shown in Fig. 14. The secondary pulse is recorded in the same way with the geophone
connected directly to the galvanometer as with the amplifier in the circuit. The time Δt between the arrival of the primary and secondary waves is practically the same in both cases.

Thus, the seismic amplifiers do not cause distortion which would explain the appearance
of secondary pulses on records of the direct wave.

The last element in the recording system which might contribute distortion is the galvanometer, which, with a frequency of 130 cps, can be considered to be a broadband element
[13]. With the presence of the secondary pulse, the apparent frequency shift of the record by
the free surface does not exceed 10-20 cps. Consequently, the galvanometer cannot cause distortion in the recorded signal.

Considering these experiments, we may conclude that the secondary pulses are not explained by distortion of the direct wave during recording.

Conversion of Parabolic Surface Motion by Geophones

as an Explanation for Secondary Pulses

It has been stated in references [11, 12] that the displacement of the free surface, with a
source buried in the medium, is approximately parabolic. It might be hypothesized that a geophone converts the low-frequency parabolic motion of the Earth's surface into two pulses, corresponding to the beginning and the end of the parabola. Let us consider this question
in more detail.

We can write the response of a geophone as [13]:

$$v'' + 2hv' + n_0^2 v = -a\xi''',$$

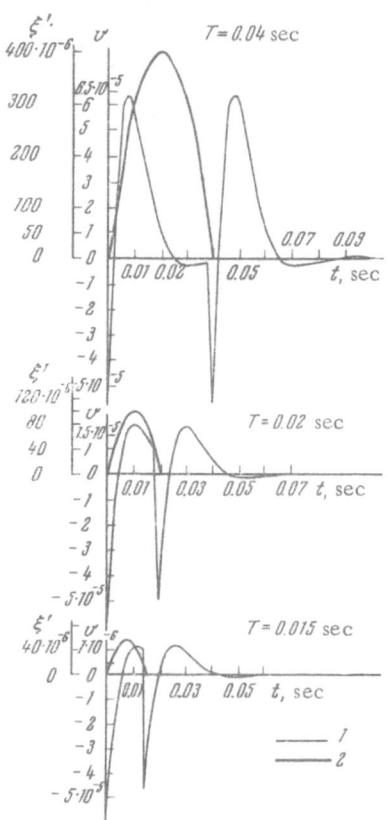

Fig. 15. Form of the signal (v-scale) at the output of an SPM-16 geophone (1) with a parabolic velocity displacement (ξ'-scale) of the geophone body (2). T is the length of the parabola in time.

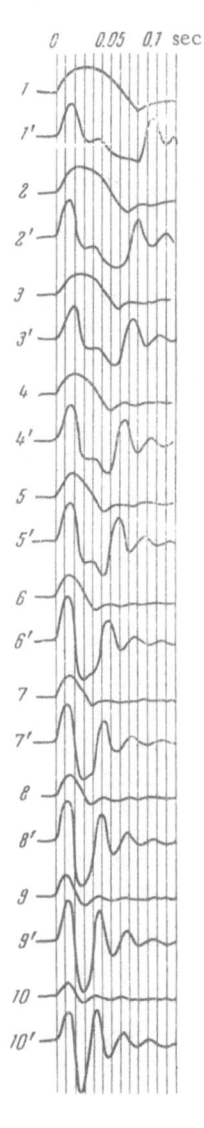

Fig. 16. Group of records of the motion of the inertial system of SPM-16 geophone (1-10) with the signal (1'-10') observed at the geophone output, obtained in modelling secondary pulses.

where v is the voltage developed at the coil of the geophone, h is the damping factor for the geophone, n_0 is the natural frequency for the resonant system, a is a constant factor in the case of a purely active suspension, and $\xi(t)$ is a function of the displacement of the geophone case.

We will take the velocity $\xi'(t)$ of the geophone coil relative to its stable mass as the input to the geophone. We assume that

$$\xi'(t) = \begin{cases} 0 & \text{for } t < 0, \\ Tt - t^2 & \text{for } 0 \leqslant t \leqslant T, \\ 0 & \text{for } t > T, \end{cases}$$

where T is the length of the parabola in time.

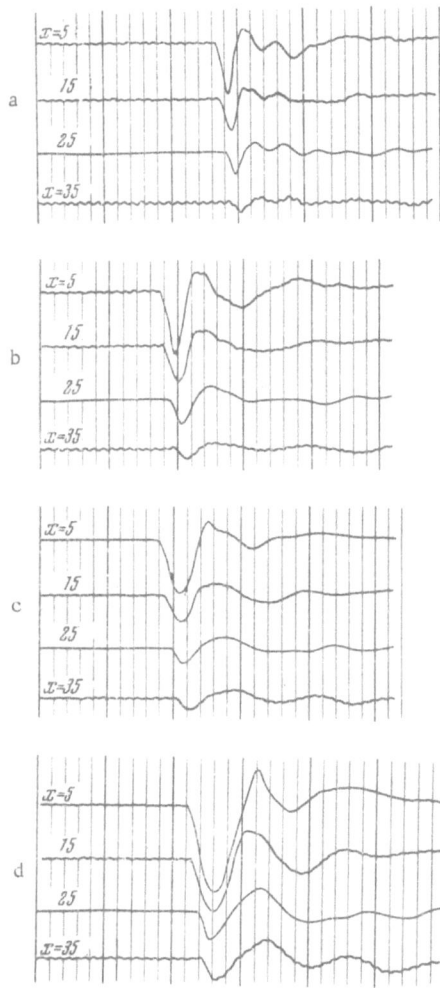

Fig. 17. Examples of the motion of the Earth's surface recorded with a VEGIK seismograph connected to a GB-IV galvanometer for various size shots at a depth of 21 m. a) Q = 0.4 kg; b) Q = 0.8 kg; c) Q = 1.2 kg; d) Q = 2.5 kg.

This expression may also be written as

$$\xi'(t) = (Tt - t^2)[\varepsilon(t) - \varepsilon(t - T)],$$

where $\varepsilon(t)$ and $\varepsilon(t - T)$ are unit step functions.

We find the third derivative of the displacement:

$$\xi'''(t) = -2[\varepsilon(t) - \varepsilon(t - T)] + (T - 2t)[\delta(t) - \delta(t - T)],$$

where $\delta(t)$ is the Dirac delta function.

Thus, the equation for the geophone output in this case is

$$v'' + 2hv' + n_c^2 v = a\{2[\varepsilon(t) - \varepsilon(t - T)] +$$

$$+ (T - 2t)[\delta(t) - \delta(t - T)]\}.$$

It has been shown that a solution to an equation of the form

$$\frac{d^2 g}{dt^2} + 2h\frac{dg}{dt} + n_0^2 g = \delta(t)$$

may be written as

$$g(t) = \frac{1}{b} e^{-ht} \sin bt,$$

where $b = \sqrt{n_0^2 - h^2}$.

Solving the equation for geophone output, considering superposition, we have

$$v = e^{-ht}\left[Tb^{-1}\sin bt - \frac{2}{b(n^2 + b^2)}(h\sin bt + b\cos bt)\right] +$$

$$+ e^{-h(t-T)}\left\{\frac{2}{b(n^2 + b^2)}\left[h\sin b(t - T) + b\cos b(t - T) + \right.\right.$$

$$\left.\left. + Tb^{-1}\sin b(t - T)\right]\right\}. \tag{1}$$

Equation (1) was used to compute the signal at the output of a geophone for a parabolic displacement of the geophone case (Fig. 15). The signal at the geophone output consists of two decaying pulses at the beginning and end of the parabola. The two pulses have the same polarity onset. With decreasing T, the pulses occur closer together and begin to overlap. The behavior is similar to the behavior of the secondary pulses as the observation point is moved from the mouth of the shot hole.

These results have been confirmed experimentally. A parabolic signal of various lengths was applied to one of the coils in an SPM-16 geophone. The signal from the other coil was used as the input to an oscillograph. A group of recorded signals at the input and output of the geo-

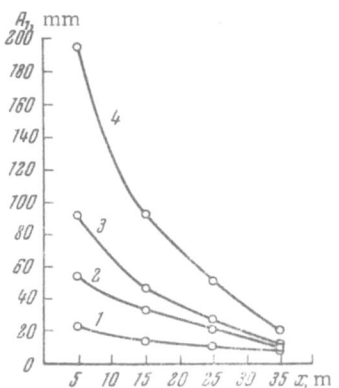

Fig. 18. Curves for the relationship of the amplitude of the first extremal in displacement to the distance from the shot hole for various charge sizes. 1) Q = 0.4 kg; 2) Q = 0.8 kg; 3) Q = 1.2 kg; 4) Q = 2.5 kg.

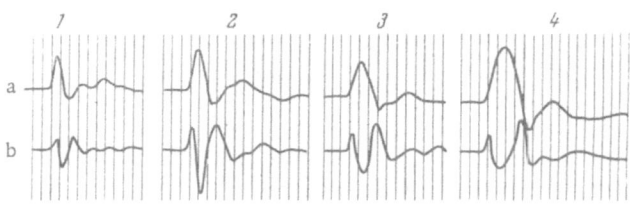

Fig. 19. Groups of records of the motion of the free surface (a) and the direct wave as recorded with an SPM-16 geophone and ShPSS recording equipment (b) for shots of various sizes at a depth of 20 m. 1) Q = 0.4 kg; 2) Q = 0.8 kg; 3) Q = 1.2 kg; 4) Q = 2.5 kg.

phone is shown in Fig. 16. Two pulses, separated in time, appear at the output of the geophone. The time Δt between the first and second pulses changes in relation to the length of the parabola. Thus, the earlier result is confirmed.

It must be noted, however, that the response function for a geophone driven by a force applied to a coil is different than its normal response function [15]. Therefore, a quantitative comparison cannot be made between these results.

Thus, the calculations and the experiment indicate that a parabolic displacement is converted into two pulses by a geophone.

Measurements were made of the motion of the free surface with a VEGIK seismograph connected to a GB-IV galvanometer, to test this hypothesis under field conditions. This system has a broad, low-frequency passband (see Fig. 1). The seismographs were located on a profile at a distance of 10 m from the mouth of the shot hole. The records of ground motion are shown in Fig. 17.

Secondary pulses are absent from the record of motion. The motion recorded with the seismograph located at the mouth of the shot hole is nearly parabolic in form, and is the same for all shot sizes. The form of the ground motion changes with distance from the shot hole; the length of the first half period and the ratio of amplitudes of the first and second extrema decrease.

Figure 18 shows curves for the relationship of the amplitude of the first extremal of the vibration to the distance from the mouth of the shot hole and to charge size. There is a nonproportional change in amplitude with distance.

At the same time, the direct wave was recorded along the same profile at the same points with SPM-16 geophones and ShPSS recording equipment. The development of the secondary pulses can be seen on these records (Fig. 19).

Thus, on the basis of these experiments, it appears that the secondary pulses on records of direct waves are the result of the conversion of a parabolic form of displacement by the recording distance, and decreasing charge size results from the change in the form of the motion of the free surface.

These investigations have shown that it is not correct to use the recorded form of the direct wave as the incident pulse when the secondary pulse is present.

Conclusions

1. Secondary pulses on records of the direct wave are formed both with shots in dry holes and with shots in water-filled holes. The difference in arrival times for the first and second pulses changes with a change in charge size and shot depth. The relationship $\Delta t = f(Q)$ may be written as $\Delta t = kQ^{1/3}$.

2. The secondary pulses have a limited range. With increasing distance from the shot point, the time delay between the first and second pulses decreases, and beyond some distance, only the direct wave is recorded. Within the medium, the secondary pulse is recorded only in the low-speed zone. The secondary pulse is not recorded below the low-speed zone.

3. Comparison of the form recorded for reflected waves on traces on which the direct wave has or does not have a secondary wave indicates that the secondary pulse has no effect on the reflection.

4. It has been shown that the secondary pulse on recorded direct waves is not the result of a pulsating gas bubble in a water medium.

5. The appearance of secondary pulses on recorded direct waves cannot be explained as the effect of distortion in the recording equipment.

6. Study of the motion of the free surface indicates that the form of the movement depends on the distance from the source and the charge size. The motion of the free surface is approximately parabolic.

7. The secondary pulse on the recorded direct wave is the result of the conversion of the parabolic motion by the geophone.

The source of the secondary pulse on records of direct waves has been explained by the results of these experiments. However, the cause of the parabolic motion of the free surface is not clear. It might be assumed that this effect is analogous to the spalling of the free surface after underground explosions [16, 17]. Experiments to test this hypothesis will be continued.

LITERATURE CITED

1. Aspects of the Dynamic Theory of the Propagation of Seismic Waves, No. 2, Leningrad State University, 1959.

2. A. M. Epinat'eva, V. V. Kuznetsov, and Yu. A. Ostrovskii, Some experimental data on the form of pulses excited by shots in wells, Izv. Akad. Nauk SSSR, Ser. Geofiz., No. 6 (1963).

3. V. V. Kuznetsov, Determination of the Average Attenuation Factors and Reflection Factors from the Spectrums and Amplitudes of Direct and Reflected Waves, Tr. Inst. Fiz. Zemli, No. 34:201 (1964).

4.   S. A. Kats and T. N. Ershova, Elements of methods for experimental determination of amplitude and phase characteristics, Prikl. Geofiz., No. 50 (1967).

5.   A. M. Epinat'eva, Secondary emissions from shots in wells, Izv. Akad. Nauk SSSR, Ser. Geofiz., No. 4 (1951).

6.   A. M. Epinat'eva, Secondary emissions in seismic observations, Izv. Akad. Nauk SSSR, Ser. Geofiz., No. 4 (1951).

7.   R. H. Cole, Underwater Explosions, IL, 1950 [current English edition: Smith, Peter (Gloucester, Mass.), Dover* (New York, N. Y.), 1965].

8.   L. L. Khudzinskii, Broadband seismic equipment ShPSS, Izv. Akad. Nauk SSSR, Ser. Geofiz., No. 4 (1963).

9.   D. P. Kirnos, B. G. Rulev, and D. A. Kharin, The VEGIK Seismograph, Tr. Inst. Fiz. Zemli, No. 6:183 (1961).

10.   E. S. Borisevich, Portable multichannel electromagnetic oscillograph, Priborostroenie, No. 10 (1956).

11.   B. G. Rulev, Similarity of wave compression for shots in soil, Prikl. Mekhanika i Teknicheskaya Fizika, No. 3 (1963).

12.   A. V. Nikolaev, Nature of the elastic wave field close to a vibrator source on a free surface, Izv. Akad. Nauk SSSR, Ser. Fiz. Zemli, No. 3 (1965).

13.   I. I. Gurvich, Seismic Exploration, Gostoptekhizdat, 1960.

14.   A. A. Kharkevich, Spectra and Analysis, Gostoptekhizdat, 1953 [English translation: Consultants Bureau, New York, 1960].

15.   L. I. Bokanenko, Determination of the frequency and phase response of electrodynamic geophones using excitation by an auxiliary coil, Izv. Akad. Nauk SSSR, Ser. Geofiz., No. 7 (1956).

16.   D. Wendle Weart, Particle motion near a nuclear detonation in hallite, Bull. Seis. Soc. Am., 52(5) (1962).

17.   J. D. Eisler and F. Chilton, Spalling of the earth's surface by underground nuclear explosions, J. Geophys. Res., 69(24) (1964).

---

* Paperback.